QUANTUM
TECHNOLOGY

Professor Gerard J Milburn has worked as a Royal Society Research Fellow at Imperial College, London, as a lecturer at the Australian National University, Canberra and he is currently Professor in Theoretical Physics at the University of Queensland. He is the author of over 100 scientific publications in the areas of laser physics, quantum optics and quantum chaos. He is also the co-author of *Quantum Optics* (1994).

Forthcoming titles in the **Frontiers of Science** series are:

Cosmic Bullets – High energy particles in astrophysics
    Roger Clay and Bruce Dawson
Drums of Heaven – The birth of gravitational astronomy
    David Blair and Geoff McNamara
Computers, Complexity and Life Terry Bossomaier and
    David Green

About the cover image

The nanomanipulator, developed at The University of North Carolina, Chapel Hill, combines the new technology of scanning force microscopy with computer techniques and virtual reality, to give a view of the surface of graphite. The image shows scratches at nanometre scale and rippling sheets caused by planes of graphite pushing upward out of the surface.

Reproduction of this image is by kind permission of the University of North Carolina Department of Computer Science.

FRONTIERS OF SCIENCE
Series editor — Paul Davies

QUANTUM TECHNOLOGY

Gerard J. Milburn

ALLEN & UNWIN

First published in 1996 by
Allen & Unwin Pty Ltd
9 Atchison Street, St Leonards, NSW 1590 Australia
Phone: (61 2) 8425 0100
Fax: (61 2) 9906 2218
E-mail: frontdesk@allen-unwin.com.au
Web: http://www.allen-unwin.com.au

National Library of Australia
Cataloguing-in-Publication entry:

Milburn, G.J. (Gerard J.).
   Quantum technology.

   Includes index.
   ISBN 1 86448 146 3.

   1. Quantum theory. 2. Quantum theory—Industrial
applications. I. Title.

530.12

Set in 11/13 pt Plantin by DOCUPRO, Sydney
Printed by Australian Print Group, Maryborough, Victoria.
10 9 8 7 6 5 4 3 2

FOREWORD

It is probably no exaggeration to say that quantum mechanics is the most successful scientific theory in history. However, in spite of the fact that the origins of the quantum theory stretch back nearly a century, it is only in recent years that the general public has become aware of the subject. Indeed, until a few years ago, the very word quantum was almost unknown outside the scientific community. Now books with 'quantum' in the title are legion.

The reason for this late surge in interest can be traced to the truly weird nature of quantum mechanical ideas; for quantum physics amounts to much more than a theory of atomic and subatomic processes. It represents nothing less than a complete transformation of our world view. Its implications for the nature of reality and the relationship between observer and observed are both subtle and profound.

A description of the world in which an object can apparently be in more than one place at the same time, in which a particle can penetrate a barrier without breaking it, in which something can be both a wave and a particle, and in which widely separated particles can cooperate in an almost psychic fashion, is bound to be both thrilling and bemusing. Niels Bohr, one of the founders of the theory, once remarked that anybody who is not shocked by quantum mechanics hasn't understood it.

For decades the sheer weirdness of the quantum world

was an obstacle to the theory being known outside the scientific community. Then in the 1970s, a number of writers recognised that the deep philosophical implications of quantum mechanics would be of considerable interest to the wider public, especially as some of the quantum mechanical concepts were of a mystical flavour. In addition, technological advances enabled certain key ideas of the theory to be tested in the laboratory for the first time, amid considerable publicity.

Although this broader interest was largely stimulated by the philosophical implications of the subject, all the while the practical applications of quantum mechanics were going from strength to strength. What the public perceived as primarily a set of revolutionary speculations about the nature of reality, professional physicists and engineers regarded as a means to make new devices and handsome profits.

In fact, quantum mechanics has always been a very practical subject. Even in the early years before the Second World War, its principles were applied to the electrical and thermal properties of metals and semiconductors. In the postwar years, the development of the transistor and the laser—two of the best-known quantum devices—heralded the information revolution.

Today we are surrounded by technology that owes its existence, directly or indirectly, to the application of quantum mechanical processes. From the humble CD player to the marvels of modern optical fibre communications, from non-drip paint to car brake-lights, from MRI hospital imaging machines to the scanning tunnelling microscope, quantum technology is now a serious money-making business.

Looking ahead to the next fifty years, quantum technology offers some breath-taking possibilities. The field of nanotechnology sets as its goal the construction of machines of molecular dimensions, with potential applications to medicine, computing and the fabrication of new and exotic materials. Already quantum technologists can trap and experiment with individual atoms, bounce atoms up and down on cunningly sculpted electromagnetic fields, produce atomic graffiti by displacing single atoms on a material surface, and image the structure of a crystal atom by atom.

These experiments probe the deep quantum regime, where Heisenberg's uncertainty principle and other aspects of quantum weirdness significantly shape the restrictions and possibilities. The common-sense world of Newtonian machines is left far behind. This is the domain of undreamt of possibilities, of microscopic circuits with novel electrical properties, of detectors so sensitive they could pick up the equivalent of the drop of a pin on the other side of the earth, of devices to make and break codes that no conventional supercomputer could touch.

Consider, for example, the bizarre properties of the quantum vacuum. Normally we envisage empty space to be just that—a featureless void. But the quantum vacuum, though devoid of ordinary particles, nevertheless seethes with ghostly activity, as so-called virtual particles spontaneously and unpredictably appear out of nowhere, only to survive fleetingly before disappearing into nothing once more.

This ubiquitous restless vacuum texture has immense implications. Cosmologists believe it may have been responsible for creating the entire universe. Stephen Hawking believes it will cause black holes to evaporate away into heat radiation. In the laboratory it shows up as slight but measurable shifts in the energy levels of atoms. More importantly, the quantum activity of the vacuum introduces a very fundamental source of noise into many practical devices. To evade this noise requires scientists to develop ways of manipulating the quantum vacuum. Advances with lasers have enabled the vacuum noise to be 'squeezed' or quietened below the natural background level, opening up the possibility of transmitting or detecting signals with unprecedented sensitivity.

Perhaps the most exciting—and most speculative—device on the quantum technology drawing board is the quantum computer, a machine that would be able to perform mathematical manipulations that are impossible, even in principle, on a conventional computer. In effect, a quantum computer could process information in many alternative realities simultaneously, and integrate them into a single real-world answer, enabling nothing less than a totally new type of mathematics to be performed.

Indeed, all quantum systems essentially exploit the fact that the quantum microworld has no single, well-defined reality, but is a ghostly amalgam of alternative universes, a hybrid world in which possible realities merge and overlap to produce a final observed actuality. Quantum technology turns this Alice-in-Wonderland realm of mind bending concepts into concrete, practical devices.

Gerard Milburn is a world-renowned theorist who works at the forefront of quantum mechanics, quantum field theory and quantum technology. His current specialism is 'atom optics'—using the quantum properties of atoms as well as those of lasers to achieve novel states of matter and hitherto unavailable methods of information flow and retrieval.

Milburn shares my excitement about the heady progress being made in laboratories around the world, turning theorists' speculations into reality. Even as Gerard was writing this book, the University of Colorado announced that it had achieved a new state of matter, a so-called Bose condensate, in which atoms of the element rubidium collectively cooperate via quantum effects to behave in some respects like a single, giant atom. Many of these bizarre new properties of matter were predicted over half a century ago by the great founders of quantum physics, such as Albert Einstein, Werner Heisenberg and Niels Bohr, theoretical physicists who developed their ideas using 'thought experiments'—imaginary situations that are logically possible but which nobody dreamed would ever become a reality. Now those dreams are being put into practice.

The nineteenth century was known as the machine age, the twentieth century will go down in history as the information age. I believe the twenty-first century will be the quantum age. Here to guide you, the reader, into that strange new world, is one of Australia's most skilled and knowledgeable quantum physicists.

Paul Davies

CONTENTS

FIGURES

ACKNOWLEDGMENTS

I would like to thank Michael Gagen for taking the time to carefully read an early draft of this book and offer many valuable suggestions and improvements. Craig Savage and Barry Sanders also offered valuable suggestions. Some of the subjects in this book lie outside the scope of my primary research interests and thus could only be written after a steep learning curve and the assistance of many people. I would particularly like to thank Richard Taylor for helping me remove some of my misunderstandings of the fascinating world of mesoscopic electronics. He also kindly provided the data in Figure 4.2 as well as the photographs of the nanostructures in Figure 4.7. I need also to thank John Panitz of the University of New Mexico for sharing some of his knowledge of scanning tunnelling microscopy. Despite the best attempts of my friends and colleagues to enlighten me, I am only too aware of the inadequacies of my understanding in these fields and hope that any errors that result in the text are not too serious.

PREFACE

The quantum theory was born amidst cultural and political ferment at the dawn of the twentieth century, and was as revolutionary and unsettling in character as contemporary developments in art, music, literature and political life. Until recently the impact of the quantum theory has been confined to scientific and, to a lesser extent, philosophical discourse. But as the century closes it is clear that the quantum message has reached the popular consciousness. A large number of books that purport to explain quantum theory have been written for a popular audience. While a few of these are quite excellent, many are based on inappropriate and misleading analogies that demonstrate little more than the authors' failure of imagination if not of understanding. It is true that the world revealed to us in the quantum theory is surprising, but to try to communicate this surprise by invoking time-worn mysticism, both western and eastern, diminishes the truth.

The aim of modern science is to reach an understanding of the world, not merely for purely aesthetic reasons, but that it may be ordered to our purpose. It is dangerous to meddle with what you do not understand, and many of the problems facing us at the close of the century reflect this fact. These problems, often blamed on too much science, usually reflect too little science. Of course our understanding is always imperfect and must be continually revised. This is no doubt true of the quantum theory as well. Yet it is clear that we understand enough of the quantum world to begin to use that knowledge in beneficial ways. A quantum technology is a technology which manipulates quantum probability amplitudes directly. This is now happening and some of the resulting technologies are described in this book.

Gerard Milburn
Brisbane 1995

CHAPTER ONE

QUANTUM ROULETTE

I admit nothing but on the faith of eyes, or at least of careful and severe examination, so that nothing is exaggerated for wonder's sake, but what I state is sound and without mixture of fables or vanity.

Francis Bacon, *The New Organon*

Like the horizon, apparent limits to the progress of modern technology fade before us as we move. The journey down to smaller and smaller scales has, until now, revealed a landscape well described by familiar concepts. But just beyond the horizon lies a new world, the coastline of which has been sighted. Within a decade high technology will be exploring the unfamiliar world of the quantum.

The world revealed to us in quantum theory is stranger than anything imagined by the wildest mystic, but we understand enough to see how it might be controlled. In this book I will take you on a journey through the unfamiliar and perplexing world of the quantum, and show you the foundations of a nascent technology and billion-dollar industry which promises to enrich our lives.

No useful description of the physical world can be given that does not account for chance and randomness. The quantum theory indicates that the universe is irreducibly random. If we are to exploit the quantum world to build a winning technology we need to understand the odds that nature is dealing us. Randomness is not a unique feature of the quantum world. All around us we see examples of apparently unpredictable and random behaviour. To appreciate what is special about the kind of randomness present at the quantum level we need to consider for a moment ordinary, everyday randomness.

We are all by now quite familiar with the notion that

1

the universe is ordered and understandable—regularities exist to such an extent that we refer to them as 'laws of nature'. It can't have always looked like this. The world of a late Pleistocene hunter must have seemed dominated more by chance than regularity. Certainly there were the regularities in the passing days and seasons, the waxing and waning of the moon, the progression of the stars across the sky. However, when the gaze of an aspiring ice-age physicist returned to the everyday world, life was not so ordered. Each day might bring some entirely new and unpredictable event: an earthquake, a hungry bear sleeping at the cave door, or perhaps the unaccountable caprice of the tribal chief. The regularity in the world must always be viewed through the obscuring haze of randomness, noise and error.

The instinct to seek order in a matrix of chaos is deep, but to try and explain everything at once is foolish and likely to lead to mysticism rather than science. The essential step in the development of western science was the discovery that the order of the world could be more clearly seen if the background noise was artificially suppressed—the discovery of the controlled experiment. The idea is to try to reduce influences that, at first sight, appear to be inessential. This idea lies at the heart of the scientific method. It is expressed very clearly in the writings of one of the earliest exponents of the method, the seventeenth-century philosopher, Francis Bacon—'I seek out and get together a kind of experiments much subtler and simpler than those which occur accidentally'.

That nature could be understood by experimental artifice is the crucial idea that distinguishes the new science of the seventeenth century from the earlier ideas derived from Greek science. In addition to enabling a new path to understanding, the experimental method indicated that wild nature could be controlled. The goal of the new science was not simply to gain understanding for the sake of it. For once the world is understood, it may be ordered to our purpose. As Bacon put it, 'For the end which this science of mind proposes is the invention not of arguments but of arts'. This pragmatic impulse to modern science is not always easy to discern in the research goals of many basic scientific programs, but it is there implicitly, even if the scientists themselves sometimes lose sight of it in the day-to-day struggle to understand the world in all its subtlety.

The most famous example of the experimental method is Galileo's alleged observations of objects falling from the Tower of Pisa. It takes only a little experience of the world to come to the opinion that not all objects fall at the same rate. It takes a considerable leap into the scientific world-view to realise that this is due to the obscuring effects of air resistance. It took the genius of Galileo to devise experiments in which the effect of air resistance could be reduced, and thus to discover that all objects are accelerated by gravity at the same rate.

However, the effects of air resistance cannot totally be eliminated, and thus there will be small variations in the measured rate of falling objects. All such variations are lumped into an 'experimental error', a small unpredictable variation in the results of the experiment caused by our inability to fully control the situation. Despite the error, however, much has been achieved—a discovery of a new law of nature. Today when we introduce our students to the scientific method we devote a great deal of time to teaching them to account correctly for errors in their results. No one should accept an empirical result unless it is accompanied by some 'margin of error'. (Why do editors often not print the margin of error when reporting statistical observations? If 45 per cent prefer candidate A and 47 per cent prefer candidate B, with a margin of error of 1 per cent, candidate A should worry. But if the margin of error is 4 per cent, nothing can be concluded at all.)

Despite the unavoidable noise and error in any real measurement, there is an instinctual belief that, were enough known about a physical system, the result of any measurement could be predicted with certainty. The triumph of physical science from the sixteenth century to the start of the twentieth century has certainly enforced a faith in the inherent predictability of the world. The primary source of this faith rests on the revelations of Newton's mechanics. Newton showed that precise mathematical relationships could be discovered behind the perplexing multiplicity of everyday phenomena, from the fall of an apple to celestial motions. The scientific legacy of Newton is Newtonian mechanics, a clockwork universe in which all matter followed a predetermined course, slowly unfolding a primordial configuration.

At the close of the nineteenth century, however, some disturbing hints began to emerge that this faith might be misplaced. The great French mathematician Henri Poincaré was the first to glimpse the jungle beyond the apparent simplicity of Newtonian mechanics. Poincaré was studying the motion of three objects undergoing mutual gravitational attractions, using powerful new geometrical tools which he had developed. He soon realised that the paths of motion must be exceedingly complex and twisted, with an infinite number of foldings and stretchings. As Poincaré himself says, 'One is struck with the complexity of this figure that I am not even attempting to draw' (Stewart 1989).

We now know that what Poincaré saw was but the tip of an iceberg. In all but the simplest and most exceptional systems, predictability is severely limited by the presence of chaos. If a system is chaotic, small errors in the initial conditions are amplified very rapidly, leading to apparently random behaviour, unrelated in any simple way to what we imagined were the starting points of the motion. But that is another story.

At about the time Poincaré was puzzling over the unimaginably complex motion of three gravitating objects, the first harbingers of a far more radical crisis for Newtonian mechanics had begun to appear. The first shot in the coming revolution was fired by the German physicist Max Planck, on 14 December 1900, the year following the publication of Poincaré's ideas. This day is the birthday of quantum mechanics.

In the last decade of the nineteenth century, many physicists were occupied with what at first sight seems a remarkably dull subject, the light emitted by hot, black balls. Hot objects emit energy as light. How hot they are determines the colour, or frequency, of the light emitted. The objective was to determine how changing the temperature changed the energy given off, and as a corollary, the frequencies emitted. The problem was the usual one: theory and experiment did not agree. This was known as 'the black body problem'.

Physical theories are the child of their time and the prevailing theoretical attempts to explain the radiation of hot objects, when Planck commenced his work, were no exception. The two great triumphs of nineteenth-century physics, electromagnetism and thermodynamics, were continuum theories—theories that regarded nature as continuous and not particulate. So strong was this belief that doubts were even raised about the validity of atoms as the ultimate building blocks of the world. If pressed, I suspect a physicist of the time might have acknowledged that, at bottom, all was atoms moving according to Newton's laws. However, such a description might have been regarded as an unnecessary extravagance. Planck adopted a more cautious attitude to the atomic hypothesis. In 1895 he wrote 'I do not intend, at this point, to enter the arena [on behalf of] the mechanistic view of nature; for that purpose, one has to carry out far reaching and, to some extent, very difficult investigations' (Mehra and Rechenberg, 1982).

It was left to Ludwig Boltzmann to attempt just these very difficult investigations by taking a radical step; Boltzmann supplemented mechanics with statistics, and in so doing laid the grounds for Planck's bold solution to the black body problem and the beginning of the end for Newtonian physics. Boltzmann's attempt to found thermodynamics on mechanics, supplemented with statistics and probability, generated enormous opposition. That Boltzmann had recourse to the atomic hypothesis was bad enough, but to introduce statistical reasoning into mechanics, the most exact of sciences, was an anathema to many of his contemporaries. In the ensuing controversy Planck took the side of Boltzmann, if somewhat ambivalently. Within six years Planck would solve the problem of black body radiation in such a way as to vindicate Boltzmann's approach and to simultaneously usher in the revolution of quantum mechanics.

The acceptance of the probabilistic ideas of Boltzmann led directly to the discovery that nature was irreducibly probabilistic. On Sunday 7 October 1900, the experimentalist Heinrich Rubens visited Planck at his home and told him of the latest results on black body radiation. Planck immediately sat down to incorporate these results into his thinking about the problem, with the result that he was able to write down a new formula. That same evening he communicated his result to Rubens on a postcard. Two days later Rubens returned to Planck. His formula agreed perfectly with the experimental observations.

So in October of 1900, Planck had the correct formula to describe black body radiation. Unfortunately he could not justify it on purely theoretical grounds. It did not take him long to find the crucial step needed to get the right result. To take that step Planck returned to Boltzmann's work, and made a decisive commitment to Boltzmann's probabilistic reasoning. If he used Boltzmann's statistical formulation of thermodynamics, he found he could derive the black body formula by making only one additional assumption: the energy of an oscillating particle is restricted to be an integer multiple of its frequency of oscillation times, a new fundamental constant now called Planck's constant. Planck had solved the problem of black body radiation. On 14 December 1900 he announced his result to a meeting of the German Physical Society in Berlin. Quantum mechanics was born.

Planck's bold hypothesis stood as a wild, exotic bloom amidst the classical landscape of nineteenth-century physics. It certainly attracted attention, even admiration, but from

Figure 1.1   A billiard-ball oscillator

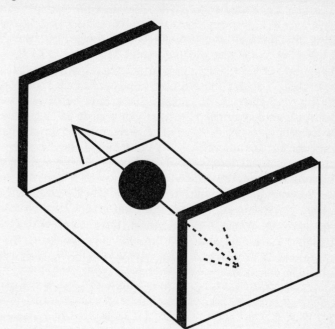

Note: A simple mechanical oscillator in which a ball bounces elastically between two impenetrable walls. This is not a simple harmonic oscillator. The period of a bouncing-ball oscillator depends on how fast the ball is moving, and thus depends on the energy of the particle. A simple harmonic oscillator has a period independent of the energy.

where did it come? While it explained the experimental results, the hypothesis itself remained unexplained. There is nothing in the physics of Newton which would require the motion of an oscillating particle to be restricted to certain energies.

We can imagine a very simple kind of oscillating particle as follows. Suppose we have a particle constrained to move only in a straight line, without friction, between two reflecting walls (see Figure 1.1). Between the walls, Newton's laws tell us the particle will continue to move, in the same direction and at constant speed. When it hits the wall it experiences a force which reverses its motion. (To keep things simple we will assume that it loses no energy at each reflection.) The particle can have any speed at all. However, if Planck's hypothesis is correct, this cannot be true. If we restrict it to have only particular energies, it cannot move at an arbitrary speed. It is as if there were some strange road rule to the effect that, on the quantum highway, you can travel at 20 km/h or 40 km/h or 60 km/h . . . but *never* at any other speed!

Twenty years would pass after Planck's address in Berlin, before a complete quantum theory, totally supplanting Newtonian mechanics, would arise to explain his hypothesis. Another sixty years would pass before technology would begin to exploit this new understanding to build quantum-well devices. The basic idea behind the quantum theory of an oscillating particle is simple enough, and I will explain it later in this chapter. The explanation turns on a new way to calculate the 'odds' for measurement results, a new probability calculus; a calculus so strange and unfamiliar that even today it generates heated debate among physicists (and others) about exactly what it means. To get to this point, however, we need to reflect a little on how ordinary odds are calculated.

Two-up with qubits (not two bits)

Sometime during the First World War, a Turkish airman, dropping bombs on assemblies of Australian soldiers, came across a small group engaged in a remarkable activity. The airman, a devout man, was most impressed as he watched each soldier periodically raise his eyes to heaven and then bow to the ground, apparently deep in contemplation of the almighty. Clearly it would have been sacrilege to attack a group so religiously occupied. The airman flew on. It was as well he did not observe too closely, for had he done so he would have seen two small coins flung into the air and fall to the dusty ground, ardently followed by the eyes of each soldier. This was not a prayer meeting but a two-up school (Sheehan and Lamotte 1985).

Two-up has been described as Australia's very own way to part a fool and his money (Simpson and Weiner 1989). As the name implies it is a game of chance played by throwing two coins. The rules are as humble as the game's origin. When played in returned servicemen's clubs on Anzac Day, two-up has simple rules which ensure even odds. When played in Australian casinos, the rules are changed to give the house an income. Two coins (pennies) are thrown into the air from a small wooden palette by the spinner, selected from the assembled crowd. The spinner aims to toss a pair of heads three times, before tossing a pair of tails or five consecutive odds (a tail and a head). If this is done the spinner is paid at 7:5, otherwise the bet is lost. Onlookers may place side bets, but here I will only discuss the game from the point of the spinner.

In one spin there are four possible outcomes, two heads (HH), two tails (TT), a head and a tail (HT) or a tail and a head (TH). How do we get from these simple facts to a calculation of the odds of a spinner tossing five consecutive odds? In order to answer this question we need to know how to assign probabilities, or odds, to the elementary events which describe the outcome of each spin. Once we have these, we need to know how to combine these elementary events to get the odds for more complex events. The mathematical theory of probability arose precisely to answer such questions about games of chance but has long since shrugged off its dubious genesis to become a respected branch of mathematics. Most people find probabilistic reasoning rather difficult, which is just as well for the continued viability of casinos and lotteries. Fortunately, we only need a couple of simple results to understand two-up.

Games of chance are probably as old as humanity, but

a mathematical theory of randomness did not arise until the Renaissance. It is rather puzzling that ancient Greek mathematics, which attained excellence in many areas, had little or nothing to contribute to the mathematics of chance. Perhaps studying games of chance was considered to be 'poor taste' in the intellectual circles of Aristotle or Archimedes, but more likely such questions didn't interest thinkers obsessed with geometry. A true mathematics of probability did not really get going until about 1660. About that time Antoine Gombault de Méré, a gifted nobleman with a penchant for gambling, asked his friend, the young Pascal, the following question: if two dice are thrown, how many tosses are needed to have at least an even chance to get a double six? Pascal's answer is recorded in a letter he wrote to another great mathematician, Fermat, and represents the birth of mathematical probability in the western science. By the way, the answer to the Chevalier's question is 24 throws for a probability of 0.491 and 25 throws for a probability of 0.505.

Pascal's answer was arrived at through a novel counting argument supplemented with one essential assumption: each result in a single roll of a die is 'equally probable'. This came to be known as Laplace's rule of insufficient reason. Given this assumption, the rest is arithmetic. To see how it works let us look at the simpler case of a toss of two coins. Four results (TT, HH, TH, HT) are possible. If each is equally probable we 'assign' the probability of one-quarter to each. Now we can calculate the probability for say an 'odd', that is a TH. Of the four possible outcomes this represents half of them, so the probability is one-half. In other words we add the fundamental probabilities for each way an odd result can occur. Given the initial assignment of probabilities, the rules of arithmetic can be applied in an almost mechanical way to answer any number of questions, but is the original assignment reasonable?

When we ask this question we are going straight to the heart of what is so puzzling about probability. The assignment of the fundamental probabilities is based on our *belief* in what constitutes a fair die or coin. Of course we can check how reasonable this belief is by taking a particular coin and throwing it many times. If it always came down heads, we would soon begin to lose confidence in our original assignment of probability. As more knowledge becomes available we need to reassign the fundamental probabilities. It seems that assigning probabilities is a curious mixture of belief and practice, of subjective and objective. It is this duality that makes probabilistic reasoning so difficult and gives philosophers endless cause for debate. It is therefore especially disturbing when probabilistic reasoning turns out to be useful in physics, which is supposed to be the objective science par excellence.

In some cases the rule of insufficient reason is itself insufficient. Common everyday usage of the word 'probability' seems to require more general rules. In some cities in Australia, the weather bureau adds to its daily forecast a statement of the 'probability of rain'. Just what does a 20 per cent probability of rain actually mean and how is this number arrived at? It is far from clear how the rule of insufficient reason could be applied here. What are the equally likely alternatives? In some cases there are just too many alternatives. For example, if you randomly select a number on the real line, what is the probability that the number selected is a rational number? There are an infinite number of rational numbers, but there are infinitely many more non-rational numbers. In this case the probability of getting a rational number is zero, which does not mean it is impossible. Then there are special events. For example, what is the probability that life will arise in a universe just like ours? Presumably this is 100 per cent, but this doesn't quite answer the question. What we are really asking for is just how unlikely is life on earth. Was it a certainty or was it a one-in-a-million chance? Assigning probabilities is a far from straightforward matter.

Despite these misgivings the rule of insufficient reason does seem quite reasonable, at least for games of chance, and we can always check it against experience. Once the basic probabilities are assigned, the rest follows from arithmetic. Until the early years of the twentieth century the probability calculus initiated by Pascal was the only game in town. After a while it begins to look as if it is the only possible way to calculate with probabilities. What a surprise, then, when experiments on matter at atomic scales suggested a new probability calculus—quantum mechanics. In quantum mechanics the probabilities of elementary events are determined in an entirely new way, using a new set of numbers, complex numbers, unknown to Laplace or Pascal. As a result, some of the most obvious rules of classical probability theory do not apply under certain circumstances. One such rule is the sum rule used to calculate the probability for an odds in a two-up game. Knowing how such a simple rule can fail takes us to the heart of the quantum theory.

Laplace's rule can be used when we have no idea what to expect for a given outcome. But in physics we know a lot about what to expect for a given arrangement of matter. If we know enough about the set-up of a two-up game—for example, we know the mass and shape of the pennies, their exact speed and orientation as they leave the skip, the effect of gravity and of the passing breeze—we might just be able to predict the outcome and clean up. On the other hand, even with the fastest super computer, we might die before we had finished the calculation for just one toss. In every experiment an exact description is impossible in practice and even unnecessary. If we want to get anywhere we include just the essential information and treat everything else as a source of randomness or noise. We can always console ourselves by continuing to think that a completely predictable description can be given.

In quantum theory, however, this classical ideal is forever abandoned. It is now known that at the atomic level nature is irreducibly random. Even if the entire experimental arrangement is completely specified, we cannot predict with certainty the results for all possible measurements we might care to make on the system. This is a deeply disturbing fact. The goal of physics has always been to provide explanations for observations. But quantum physics teaches us that this goal can be approached but not reached. Fortunately the theory does at least tell us how to calculate the probabilities for a particular observation even if in a particular run of the experiment we cannot predict with certainty what the result will be. We cannot avoid talking about probabilities if we are to describe the universe as it is.

Quantum theory is expressed in terms of probabilities for outcomes to measurements. That is not surprising. As we have seen, to get a realistic and practical description of the physical world we must put up with a certain amount of randomness. However, the rules for reckoning quantum odds are nothing like the obvious rules used in two-up. Quantum two-up is the fairest game in the land. I am going to describe a quantum two-up game played with light, but to begin we need to pause and ask what is light? The answer to this question at the end of the nineteenth century, and the one you are likely to get today if you ask a high school physics student, is: light is a wave. This is not correct. Five years after Planck introduced the quantum hypothesis to explain observations on black body radiation, another exotic quantum bloom appeared in the garden. On 9 June 1905, a paper appeared in *Annalen der Physik* with the title, 'On a Heuristic point of view about the Creation and Conversion of Light'. The author was Albert Einstein. In a letter to a colleague he described the contents of this paper as 'very revolutionary'. A considerable under-statement.

Consider for a moment the matter of air pollution monitoring. Many governments now require that the quality of our air be monitored closely. In some cities one needs no more sophisticated a monitoring device than a human nose. In more fortunate cities it is as well to find more sensitive devices to spot the trend before it is too late. One of the many villains in the cast are the oxides of nitrogen.

A standard method to monitor emissions of these substances is to continually sample a small amount of the gas going up a factory chimney. When oxides of nitrogen react with ozone, they emit light. But not much. By using the remarkable property of certain metals to develop a charge when exposed to light, the little bursts of light produced may be detected electrically. The explanation of this property was the subject of Einstein's paper.

If we take a metal plate and expose it to light it will acquire a positive charge. The explanation had been worked out six years before Einstein's paper of 1905. The light causes negatively charged electrons to be ejected from the surface. If the plate is electrically neutral to begin, losing a few electrons leaves it with an unbalanced positive charge. This seems straightforward enough until you examine things a little more closely. If you carefully control the wavelength of the light falling on the metal you find that above a certain wavelength, the light has no effect. The plate remains neutral. If you set up a little experiment to measure the speed of the ejected electrons you discover a very odd result. The maximum speed of ejected electrons, if any are ejected at all, does not depend on the intensity of the light. Increasing the intensity only increases the number of electrons ejected per second.

Both of these facts are inconsistent with the notion that light is a wave. Imagine a boat tied to a mooring, rocking up and down as waves pass by. If the height, or intensity, of the wave is big enough, the boat may break free of its moorings. If light were a wave, increasing its intensity should eventually supply sufficient energy to wrest the most tightly moored electron free of metal atoms in the surface. Yet that is not what happens. If the wavelength is not right, the electrons will remain bound to the surface of the metal, no matter how intense the light. Yet clearly, if the light supplies sufficient energy, the electrons must eventually break free. If increasing the frequency of the light causes the electrons to be emitted then there must be some relationship between the energy carried by light and its frequency. That was Einstein's insight (and the one that gained him the Nobel Prize). He postulated that the energy carried by light was proportional to its frequency, and that the constant of proportionality was exactly the same constant that Planck used in his theory of black body radiators. Furthermore, according to Einstein's explanation, the best way to picture what is going on is to regard light as made up of little particles, photons, each with energy proportional to the frequency of the light. In this way we see why increasing the intensity increases the number of electrons ejected. For each photon of the right energy one electron can be ejected. Increasing the intensity increases the number of photons and thus increases the number of electrons emitted.

Thus began the quantum theory of light, which was not to achieve a full expression until the middle of the twentieth century, long after a quantum theory of matter was accepted. (Why? Light moves so fast that any theory of light must take into account the theory of relativity. A relativistic quantum theory was not so easy to come by.) We don't usually need the quantum theory to describe light, but if we use light of very low intensity a quantum description is necessary. To describe light of different intensity, we use light of differing numbers of photons. To describe light of different colours we use photons of different energies. Before Einstein, all experiments with light were adequately explained by picturing light as a wave. Einstein used a particle picture to describe the observed effect of light on matter. So is light a particle? No, it is not. Neither is it a wave. However, if we are to communicate with each other we must use familiar concepts, such as particle or wave, to describe the outcome of observation. As long as we only use our picture in the context of the observation we are trying to describe we won't get into trouble. That at least was the interpretation given to quantum theory by the Danish physicist Niels Bohr. It was an interpretation that was never accepted by Einstein. He continued to believe that it was possible to say just what light really was, without some intellectual tightrope act in which properties depend on the act of observation. It is a great irony that one of the originators of the quantum theory should have remained, until his death, the most eloquent opponent of Bohr's interpretation of the new theory. Today most physicists accept that Einstein's hope was misplaced. But scientific (unlike religious) explanations are contingent and open to revision. One day our assessment of the Bohr–Einstein debate could be very different. So now, back to gambling.

To make a quantum two-up game with light we will use a device called a beam splitter. This is essentially a half-silvered mirror; half the light goes through and half gets reflected. What happens if we use light of such low intensity that it has only one photon? When the photon encounters the beam splitter it can either be reflected or transmitted. Let us call a photon transmitted a head (H) and a photon reflected a tail (T). If we repeat the experiment many times (many tosses of the coin) we see that roughly half go through and half get reflected. So the behaviour of any given photon is as random as a coin toss. Even if we know everything there is to know about the light we use, its intensity, colour, polarisation, we cannot predict whether any individual photon will be reflected or transmitted. The best we can do is give the odds (even in this case).

To make a quantum two-up game we must somehow toss two photons simultaneously. In our experiment we can do this by ensuring two photons arrive simultaneously at the beam splitter. So now direct two beams onto the beam splitter (see Figure 1.2). If we use a special light source, which I will describe below, we can ensure that each beam contains one photon and they arrive at the mirror together. Now there are four possible ways the photons can get to a final detector: both photons can be transmitted, both photons can be reflected, and two ways in which one photon is reflected and one is transmitted. The first two possibilities result in a single photon being detected at each detector

Figure 1.2 Two-up with two photons

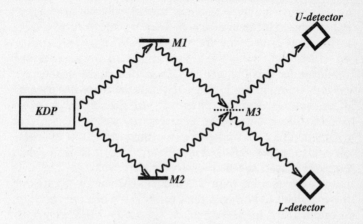

Note: The experimental realisation of a coin toss with two photons, which travel along the paths shown. The source of the light, labelled KDP, produces two photons simultaneously. These photons are directed by two mirrors, M1 and M2, onto half-silvered mirror, M3. At this mirror a photon has a 50/50 chance to be reflected or transmitted. Finally, two photon detectors are placed to intercept all the light from each of the two output beams. Contrary to classical expectations we *never* see a single count simultaneously at both detectors.

---

simultaneously. The last two possibilities result in two photons arriving simultaneously at one detector. In the end we can only count how many photons arrive at the two output detectors, but we have no way of knowing if it arrived there by reflection or transmission. To see the connection with two-up more clearly, label the two output detectors with the symbols T and H. By analogy with the two-up game we would expect to see one photon simulta- neously at each output (TH) twice as often as seeing both photons in the same output only, HH or TT (one reflected and one transmitted). This experiment was performed at the University of Rochester, New York, in 1987 (Hong et al. 1987). The result—one *never* sees a single photon in each output simultaneously. In the language of the experiment, the coincidence rate is zero. In the language of a quantum two-up game the probability of an odd is zero.

If we were to see this happen in a two-up game we would, quite sensibly, ask to take a look at the pennies, or perhaps question the honesty (but not the manual dexterity) of the spinner. But one photon is just like any other. If we were to block one of the twin beams, we would see that a photon from the other beam would still have an equal chance of being reflected or transmitted. It is only when both are tossed together that we see something strange. Is there some cosmic spinner, of dubious honesty, controlling the toss? Has God rigged the game? The correct explanation is much more interesting.

The quantum theory explains this result quite easily, by giving a new way in which probabilities are assigned. The rule is simple. What is hard to swallow is that the universe should behave this way. In a quantum description probabilities are not the primary object. The quantum theory gives the probability *amplitude* for an event. The probability is obtained by squaring the amplitude. The rules for combining amplitudes are the same as the rules for combining probabilities. If an event can happen in two indistinguishable ways, add the probability amplitudes, then square to get the probability. Now comes the essential point—probability amplitudes, unlike probabilities, do not have to be positive numbers. A count of a single photon at each detector can occur in two indistinguishable ways: both photons can be transmitted or both photons can be reflected.

The quantum theory for this experiment shows that the amplitude for both photons to be reflected is $1/2$. however the amplitude for both photons to be transmitted is $-1/2$. As the photons are identical there is no way of knowing which is which after the beam splitter, so the amplitude to have both photons transmitted or both photons reflected is the sum of these two amplitudes, which is zero. We never detect a single photon at each detector simultaneously. In standard statistics probabilities add; in quantum statistics probability amplitudes add and thus may completely cancel. In two-up with quantum objects there are only two equally likely outcomes, HH or TT. We might as well just toss a single coin. A casino will never make any money on quantum two-up.

The transmission of a single photon through a beam splitter is just as random as the toss of a single coin. When a single coin is tossed there are two possible outcomes, heads or tails. When a photon passes through a beam splitter there are two outcomes, it is either reflected or transmitted. An experiment with just two outcomes is the elemental random event. It is convenient to have a universal way of labelling the outcomes of an elemental random event. This is done by assigning the numbers 1 and 0 to each of the two results. These numbers are called binary digits or 'bits'. A single coin toss is a one-bit game as we need only one binary digit to describe a particular outcome. Two-up is a two-bit game. Quantum two-up, however, is a two-bit game with a difference. To capture this difference a new word, 'qubit', has been coined to describe the outcome of the elementary quantum random event. Qubits will be discussed in much more detail in chapters 5 and 6.

The experiment with two photons at a beam splitter has been done. How did they arrange to get two photons simultaneously in separate beams? By using the remarkable properties of a carefully cut crystal of potassium dihydrogen phosphate (KDP for short). When intense green light is directed into such a crystal, it generates two diverging beams of red light. It works by destroying one photon at the higher frequency (the green) and simultaneously creating two photons at the lower frequency (the red). If you don't drive the crystal too hard, you can get one red photon in each beam with high probability. I have told you that the quantum theory gives us probability amplitudes from which the actual probabilities are determined by squaring. How to get the probability amplitudes in the first place is a bit harder to explain. However, in order to understand the examples in this book you only need to remember that if an event can happen in two ways, each indistinguishable, then the amplitude for the event is the sum of the amplitude for each way considered separately. I will call this Feynman's rule. (It appears in a very good set of physics books written by the late theoretical physicist, Richard Feynman, Feynman et al 1965.) We will however need a couple of other rules. If two-up is not your game, how about a little billiards?

Quantum billiards

In chapter 4 we will consider a real game of quantum billiards, with electrons for balls. Here I will discuss a particularly dull billiard table—a straight line with reflecting ends as in figure 1.1. This may seem a rather artificial example but a similar device can be built by etching structures in semiconductor chips. In this example a small particle is bouncing backwards and forwards in a straight line between two reflecting walls. If the particle is moving too fast it may knock a hole in the walls and go straight through, the walls and the particles being rather the worse for the experience. We don't want that, so we will assume the particle doesn't move so fast.

The ball may be moving to the left or the right. We won't distinguish the two cases. We will assume that we observe the position of the ball by taking a snapshot at different times. When we look at the photograph, we cannot determine which direction a particle was moving. In order to simplify things, I will only consider the case where the ball always moves at the same speed. So here is the model—a little ball moving, either to the left or to the right, at some predetermined speed, and reflected without any change in speed from two fixed points. We don't need to know much about Newtonian mechanics to describe this system! If we take a snapshot at random times we are no more likely to see a ball in the middle than at either end, or indeed anywhere else for that matter. If we superposed all the snaps we would conclude a ball was equally likely to be found anywhere.

If we do this same experiment with quantum particles, such as electrons, the snapshots show a completely different ball game. We *never* see a ball at either end. Indeed in most cases we see the ball in the middle. There is a greater chance to find a quantum billiard in the middle of the box and no chance at all to find it at either end. What are the quantum rules to explain this?

We already have the rule to explain why we never see a ball at either end. To calculate probabilities we must first assign a probability amplitude. If we can't distinguish two possibilities we first add the amplitudes, then square, to find the probability. In the billiard game there are always two possibilities which our snapshot can't distinguish: a ball found at any position can be travelling either to the left or the right. To determine the probability to find a ball at some position in the snapshot we must first add the amplitude for it to be found at that position and travelling to the left to the amplitude for it to be found at that position and travelling to the right. In some cases these amplitudes may cancel exactly—just as in the game of quantum two-up. This is precisely what happens at either end; the amplitude for a left-going ball exactly cancels that for a right-going ball, so there is no probability of finding the ball at either end. In the middle, however, the two amplitudes add together, so there is a large probability to find a ball in the middle. Everywhere else is intermediate between these two extremes, that is there is only a partial cancellation of amplitudes for left-going and right-going balls.

How are we to assign the correct amplitudes? Clearly the amplitude at each position is different and it must also be different for left-going and right-going particles. As we expect complete cancellation at either end, let us assign

Figure 1.3   Quantum mechanics of a billiard-ball oscillator

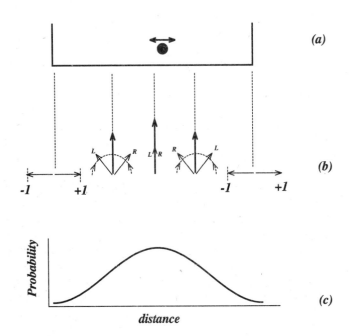

Note: A diagrammatic representation of the quantum probability calculus. A ball is bouncing between two impenetrable plates, (a). The probability to find the ball, at a fixed time, at some point between the plates is determined by a probability amplitude arrow. This arrow is shown in (b) as the sum of two probability arrows for the two ways, either travelling to the left or the right, in which a ball can be found at a particular position. At each wall the amplitudes for a left- and a right-going ball cancel exactly. In between the amplitudes for left- and right-going balls are rotated, in opposite directions, by an angle that depends on the momentum of the ball. At the centre, the rotation is such that there is maximum reinforcement between the amplitudes for left- and right-going particles, so here the probability to find the ball is a maximum, as seen in (c). We are only allowed to have momenta (or speeds) which enable the probability amplitude arrows for left- and right-going motion to rotate back to total cancellation at the opposite wall. This restricts the momenta and thus restricts the energy the particle can have.

+1 as the amplitude for a particle to be found at the left-hand edge and travelling to the right, while the amplitude to be found at the left-hand edge and travelling to the left is −1. Now, how do we assign the amplitudes for all the points in between?

Let us represent the numbers ±1 as arrows, one directed from the 0 to +1 and the other directed from 0 to −1 as in Figure 1.3. To assign the amplitudes for all the positions in between we let these arrows rotate with their tails pinned together. Here is the first rule: if a particle is moving to the right, rotate the arrow anticlockwise, if the particle is moving to the left, rotate the arrow clockwise. How far we rotate the arrow depends on two things; how far away from the edge we move and the speed of the particle. Here is the second rule: rotate each arrow through an angle proportional to the momentum of the particle (the product of the particle's mass and its velocity) multiplied by its distance from the edge. This means if we double the speed of the particle the arrow rotates twice as fast.

How do we get the actual probability from these rules? We add the arrows together to get a 'total arrow', see Figure 1.3. The probability is proportional to the square of the length of the resulting arrow. So if the arrows point in different directions, as they do at each edge, they cancel exactly to give zero probability. If the arrows point in the same direction the resulting arrow is twice the length of each and the probability is not zero. It is easy to see that the maximum probability will occur in just this way, with the arrows pointing in the same direction. So this must correspond to a particle located in the midway between the walls. This is how it looks: start at the left-hand edge with two arrows pointing in opposite directions, move along the box towards the right and rotate each arrow in opposite directions until at the middle of the box they are pointing in the same direction, continue on towards the right until we get to the other edge, now the arrows must point in opposite directions in order to cancel again.

Of course it is possible that there will be many rotations as we move along the box. All we require is that the arrows manage to cancel again when we finally reach the right-hand edge. The minimum angle we can move the arrow and get cancellation at either edge is 180 degrees. But wait, the angle through which we move the arrow depends on the momentum. For a box of fixed length, a particular speed may not get us all the way through 180 degrees. Here is an astonishing conclusion: not every speed is allowed, only those speeds are allowed which cause the arrow to rotate sufficiently to get cancellation by the time we have moved along the full length of the box. The quantum rules must restrict the allowed speeds of the particle and thus the allowed energies! This is precisely the assumption made by Planck.

What is the minimum speed allowed? This will be the speed which gets the arrows through 180 degrees—exactly— as we pass from one side of the box to the other. What this minimum speed actually is depends on the size of the box and the precise value of the constant of proportionality between angle rotated and speed of the particle. The important point is that *the minimum speed is not zero*. In quantum mechanics the energies the particle can have are restricted, and the lowest energy state does not correspond to a stationary particle as it does in classical mechanics. A particle confined to a box *must* move, it cannot be at rest, and if it moves it can only move with particular energies. This is the explanation for Planck's hypothesis. We now see that this strange quantum speed-limit is a direct result of the peculiar way in which probabilities are calculated.

You should not picture the arrows representing the probability amplitudes as somehow a *property* of the particle, in the same we regard, say, mass as the property of a particle. The arrow is no more a property of the particle than is the probability for a coin to come up heads or tails a property of the coin. Probabilities refer to the outcomes of observations on the particle. What we actually see, heads or tails, is a property of the coin, but the probability is not. We never observe a probability, we observe heads or tails, or a photon transmitted or reflected. Another reason we cannot regard probabilities as properties of particles is that probabilities have a habit of changing depending on what is in our heads. For example, if we toss a coin but do not look at the outcome, the probability of a head is %50. If we do look and find a tail, the probability of a seeing a head is now %0! Probability amplitudes like ordinary probabilities are simply numbers we assign to describe our expectations for observations. Of course, how we assign these numbers depends on the precise nature of the system under observation, but we never observe a probability or a probability amplitude.

Why are the probabilities for observations on quantum systems calculated from probability amplitudes? Is there some principle, some simply stated, self-evident idea from which the rules of quantum theory can be derived? At a recent meeting of theoretical physicists, held in the delightful Santa Fe estate of Sol y Sombra, the American physicist John Wheeler asked just this question. It is a question he has been asking at such gatherings for many years. It certainly livens up the discussions at coffee and lunch breaks, but no one has any idea what the answer is, or indeed if the question itself has any cogency.

Undeterred, John Wheeler himself has proposed an astonishing answer: '. . . all things physical are information theoretic in origin' (Wheeler 1989). John Wheeler is asking us to take as primary the statistical character of the quantum theory. His call is slowly acquiring adherents. A number of theoretical physicists have started framing quantum theory in the language of statistics with surprising results. My own group at the University of Queensland, in collaboration with Carlton Caves at the Center for Advanced Studies, University of New Mexico, and Sam Braunstein at the

Weizmann Institute in Israel, have recently shown how the uncertainty principle may be cast in the language of parameter estimation—a classical problem in statistics. This has given us new uncertainty relations, including relativistic uncertainty relations. Caves and Braunstein, building on the earlier work of Bill Wooters, have shown that considerable insight into the quantum theory is gained by recasting its fundamental formulae in terms of statistical distance—a measure of how well experiments can distinguish different states of systems. It is too early to say where these new ideas will lead, but I strongly suspect we see here the beginnings of John Wheeler's suggestion: 'Tomorrow we will have learned to understand and express all laws of physics in the language of information' (Wheeler 1989).

Quantum theory suggests to us that the physical universe is very surprising and difficult to picture. Despite these difficulties the success of the quantum theory rests on how well it fits experimental observations. Is there direct experimental evidence for quantum restriction of the energies of a confined particle? Yes there is, and not surprisingly it occurs in very small integrated electronic circuits. One example of such an experiment is based on the quantum point contact (QPC).

The electrons inside the layered structures of composite semiconductors behave like a gas, moving every which way, colliding with each other, with the boundaries of the semiconductor, and with impurities and defects in the crystal. It is possible, by careful construction, to confine the electron gas to a two-dimensional layer called a 2 dimensional electron gas or 2DEG for short. Above this layer are deposited metal electrodes (see Figure 4.1). If a voltage is applied to the metal, it can exclude electrons from the layer beneath, effectively dividing the two-dimensional electron gas into two separate regions. This arrangement is called a 'gate'. If we now make a very small gap in the gate, a split-gate, there is a narrow channel, through which electrons can pass, connecting the two electron gas regions. Effectively we have two reservoirs of electrons, moving at random, connected by a narrow channel. By applying different voltages to the metal electrodes we can change the effective width of the channel.

The channel is very small, and very few electrons can pass through it. On the way thorough an electron bounces ballistically between the walls of the channel, but it has very little chance of suffering a collision with anything else. If we were to view the channel end-on it looks just like the one-dimensional box discussed earlier. Motion in this transverse direction must be confined, and is subject to the laws of quantum mechanics. This means that the energy of the electron, associated with motion transverse to the channel, must be restricted to particular values. This device is known as the quantum point contact (QPC) as, from a macroscopic point of view, it appears as if we have two electron sources joined by a single point conductor (see chapter 4 for more details).

The objective now is to force electrons through this narrow channel. The electrons in the 2DEG have a range of energies, but only those electrons which have an energy matching one of the allowed energies for the channel can pass through the constriction. If an electron does not enter the channel, the current down the channel is zero, or equivalently the resistance of the channel is infinite. However, if the energy in the transverse direction is just right, the electron happily bounces down the channel! If we were to plot the current of the circuit versus the width of the channel, the curve would look like a series of steps—flat regions punctuated by small jumps in the current at the particular voltages corresponding to an electron getting just the right amount of energy to go down the channel. This is very different behaviour from that predicted by Ohm's law. Electronics at the quantum level requires a whole new set of laws.

The first QPC was demonstrated in 1988. The technique has since been considerably refined. The QPC provides clear evidence that the rules for calculating probabilities in quantum theory are indeed correct, but more importantly it begins to suggest a revolutionary new kind of electronics governed not by the usual rules of electronics, such as Ohm's law, but by the well-understood, though conceptually bewildering, rules of the quantum.

CHAPTER TWO

ATOMIC CALLIGRAPHY

In the film *Forrest Gump* the lead actor, Tom Hanks, meets, talks and shakes hands with a dead American president. Or so it appeared. We all know that Kennedy was not resurrected to star in a Hollywood film, so we guess that what we saw was a bit of movie magic, conjured up with a little help from computer image processing. In this age of digital image manipulation we cannot trust any image, moving or still. Digital image construction is not confined to cinematic technology. It now plays a central role in a wide variety of scientific and engineering applications. When scientific measurements were presented in tables of data, interpretation required a deep understanding of the underlying physical processes. Now a wide range of computational tools can be brought to bear on the presentation and analysis of scientific data. Often this involves sophisticated graphics tools and even animation. Such tools have been pivotal to the development of new kinds of microscopes, particularly scanning probe microscopes. The technology of digital image manipulation was the midwife to a new quantum technology, the scanning tunnelling microscope (Binnig and Rohrer 1985).

Scientific research is a chaotic business, stumbling along amidst red herrings, errors and truly creative insights. Great scientific breakthroughs are rarely the work of a single researcher plodding slowly but inexorably towards some final goal. The crucial idea behind the breakthrough may surface a number of times, in different places, only to sink again beneath the babble of an endless scientific discourse. Great ideas have their day, a number in fact, before eventually someone puts it all together. One of the best descriptions of the process of research was given by two IBM physicists, Gerd Binnig and Heinrich Rohrer, on the occasion of their Nobel Prize presentation address. In describing how they built the Scanning Tunnelling Microscope (STM), Binnig and Rohrer (1987) stated: 'During this development period, we created and were granted the necessary elbow-room to dream, to explore and to make and correct mistakes'. The product of Binnig and Rohrer's dreams was the scanning tunnelling microscope, the fundamental principle of which takes us to the heart of the most surprising phenomenon in the quantum world—quantum mechanical tunnelling.

The STM enables single atoms on a surface to be imaged, providing an astonishing view of matter at atomic scales. The same device can be used to manipulate single atoms, slide them around on a surface, and build atomic scale devices an atom at a time. The implications for device miniaturisation should be apparent. Already the STM has led to a multimillion-dollar industry, with applications ranging from biotechnology to materials engineering (*Business Week* 1993). I will return to some of these applications later in this chapter, but first I want to explain the physical principle that lies behind this remarkable device. What is quantum tunnelling?

Quantum tunnelling

We all know that an atom is made up of electrons constrained to orbit a nuclear core by strong electrical forces. Inside a metal, however, atoms form a crystal lattice, also held together by electrical forces. The outermost electrons on each atom wander from one atom to the next to such an extent that they may almost be regarded as completely free (see chapter 4). These electrons form a gas of little charge carriers, moving through a lattice of positively charged atomic cores. In the gas, electrons dash about colliding with other electrons, but more importantly, colliding with the atomic cores of the lattice. In such collisions, an electron gives up a little of its energy.

As the atomic cores are bound strongly to each other, a collision with an electron can cause an atomic core to start a very small vibration. In fact, at a non-zero temperature, the atomic cores are all vibrating a little anyway, and an electron can not only lose energy when it collides with an atomic core but can gain energy as well. On average the electrons neither gain nor lose energy, but reach an equilibrium. An individual electron, however, may have slightly more or less than the average energy.

It is the existence of the electron gas that makes a metal a conductor. If we connect a wire across the terminals of a battery, electrons accumulated on the negative terminal of the battery will repel electrons in the wire. On top of the random motion of electrons there will now be a net drift away from the negative terminal and towards the positive terminal. However, an individual electron does not move all the way down the wire. In fact it will not move very far at all before undergoing a collision with an atomic core, and losing the extra energy it has acquired from the battery, but on average there is a net flow down the metal.

At the surface of a metal the electron gas is confined by electrical attraction to the positive atomic cores. The surface of a metal is the coastline of an electronic sea. This is a bit like the way the gravitational attraction of the Earth keeps our atmosphere from drifting off into space. An individual electron may rise higher or lower on the shoreline depending on its energy, but no electron has sufficient energy to penetrate the barrier of electrical forces and leave the metal. That at least is the classical view of what happens. But electrons are quantum particles, and as we shall see, the rules are different.

Consider a ball, released from some height and bouncing vertically on a smooth surface. When released, it begins

Figure 2.1   Quantum mechanics for bouncing balls

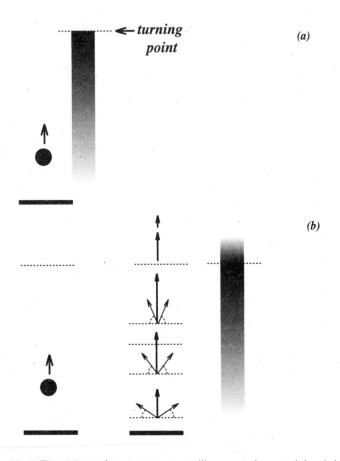

Note: The idea of quantum tunnelling can be explained by considering a bouncing ball. The top figure, (a), shows the classical picture. A ball is represented as bouncing vertically from a fixed surface. A snapshot of the ball, taken at random times, will most often show a ball towards the top of the trajectory because here the ball is travelling slowly. The grey-scale bar on the right indicates the higher probability to find a ball at the top of a bounce. No ball is ever found above the classical turning point, which is the height from which the ball was dropped.

The quantum picture is shown in (b). A measurement of the location of a ball at random times looks similar to the classical case but now some balls are found beyond the classical turning point. The measurement only tells us the height of a ball, but we cannot know if it was falling or rising. There are two indistinguishable ways in which a ball can be found at a particular height. To calculate the probability of finding a ball at a given height we must add the arrow to find a ball at that height and travelling down, to the arrow for a ball at the same height and travelling up. The probability of finding the ball at that height is then proportional to the square of the length of the resulting arrow. At the bottom of the trajectory, where the ball hits the surface, we have a situation exactly like the ball bouncing between two fixed walls discussed in chapter 1. Thus at the surface we expect the amplitude for a ball to be travelling down to cancel exactly the amplitude for a ball to be travelling up, and thus the probability to find a ball at the surface is zero. As we move away from the surface, we must rotate the arrows, corresponding to up and down motion, in opposite directions. The amount by which we rotate depends on the speed of the particle at that height. As we move up from the surface, the speed of a ball must be decreasing, so we rotate the arrows by a smaller and smaller amount at each step. We expect that by the time we get to the top of the trajectory, the arrows are hardly rotating at all and that here they should add to give a maximum probability. This is the turning point for the classical bouncing ball, but unlike the turning point at the surface, here the arrows do not cancel, on the contrary at the turning point they add more than at any other point on the trajectory. Now we must use another rule of the quantum theory: *probability amplitudes cannot discontinuously go to zero*. If a quantum particle is in a region forbidden to it by classical mechanics, the arrows stop rotating and begin to shrink. In fact they shrink very rapidly as we move beyond the turning point. But, for accurate measurements, we always find a few particles higher than they should be found in a Newtonian description.

---

to accelerate, due to gravity, towards the surface. At the surface it experiences a short, sharp force which reverses its direction. It then begins to rise, all the while being decelerated by gravity. If no energy was lost in the collision with the surface, the ball will rise no further than the height from which it was released, at which point it momentarily comes to rest, before beginning its fall all over again. In this case we say the collision with the surface is elastic. Usually, however, a little bit of energy is lost from the ball during the collision and the ball does not rise quite so high on each successive bounce and we say the collision with the surface is inelastic. To simplify matters I am going to assume that the collision with the surface is elastic.

The bouncing ball is a kind of oscillating particle; it rises and falls over and over again. How might we give a quantum description of this oscillator? Recalling the discussion in chapter 1, we expect that while the energy of a classical bouncing ball can be anything at all, in a quantum description the energies must be restricted. This is just a generalisation of Planck's hypothesis for simple oscillators. (A simple oscillator is one for which the period of the motion is fixed and is independent of the energy. The bouncing ball oscillator is not simple, as the period of the motion depends on the initial height of the ball, and thus depends on the energy of the ball.) This restriction of energies is already a surprise, but there is an even bigger surprise.

Suppose we have a bouncing ball, with one of the energies allowed by quantum mechanics, and further suppose we take a set of photographs of the ball at random times. What do we see? Let us first consider this from the perspective of Newtonian mechanics. The ball is constantly changing its speed as it falls. At the top of its trajectory it is momentarily stationary, while at the bottom it is travelling with its maximum speed. This means that if we superpose all our photographs, taken at random times, we expect to see the ball much more often towards the top of its bounce than at the bottom (see Figure 2.1). This is simply because the ball is travelling more slowly towards the top of its trajectory than at the bottom. As the ball has a fixed energy, and the collisions are elastic, the ball will always rise to the same height in each period. A careful examination of the photographs would show that there is no case in which a ball was ever found to go higher than this maximum height. This is no longer true for a bouncing quantum ball.

In the quantum world there are some photographs in which a ball is found higher than allowed in Newtonian mechanics! Somehow quantum bouncing balls can go where no classical ball is allowed to go. This phenomenon is called quantum mechanical tunnelling. The photographs show that there is a small, but non-zero, probability to find a quantum ball higher than the classical turning point. The number found beyond the maximum classical height, however, decreases very rapidly with distance. This rapid decrease is important. It is what makes a scanning tunnelling microscope such a sensitive instrument. The quantum description of a bouncing ball is given in Figure 2.1.

In classical mechanics, the turning point represents an impenetrable barrier. To go beyond this point would violate conservation of energy. In quantum mechanics the classical barrier is just a bit leaky. It is as if, every now and then, a quantum ball has 'tunnelled' through the wall separating the allowed classical region from the forbidden zone. Quantum particles seem decidedly lawless, but at least we can calculate the probability of a crime. Has anyone done an experiment like this? Not yet, at least not in exactly this form, with quantum particles falling under the action of gravity. But it will soon be possible; in chapter 3, I describe an experiment, the 'atom trampoline', done in 1993, with atoms bouncing from a fixed surface. Now, however, let us return to the story of Binnig and Rohrer and the development of the scanning tunnelling microscope.

A new technology is born

Electrons are very small, with a tiny mass, so it is not so

easy to arrange for one to fall and bounce from a surface. The scale of the bounces would be just too small to see. However, they are charged particles and thus subject to electrical forces which are much stronger than gravitational forces. Now suppose an electron in the surface of a metal starts off in a direction perpendicular to the surface. It won't get far before the strong electrical forces due to the atomic cores in the metal crystal pull it back into the electron sea. It is as if a small drop from the electron sea momentarily leapt from the surface only to fall back due to electrical attraction.

It is easy to see that in a classical description no electron can get beyond the surface to an extent that one could observe. But in quantum mechanics it can go a little further than classical mechanics permits. Every now and then an electron can tunnel into the classical forbidden zone. Suppose we put a small conducting tip very close to the surface, just outside the classically allowed region. If an electron does manage to tunnel to the forbidden zone it will now find itself in another electron sea inside the tip: it has escaped and, if the circuit is closed by connecting the tip through a wire to the metal, it appears as a small current fluctuation. This is known as 'vacuum' tunnelling as it seems that the electron has tunnelled through a vacuum between the surface of the metal and the conducting tip. It is precisely this effect that Binnig and Rohrer realised could be used to build a totally new microscope powerful enough to see single atoms.

The development of the scanning tunnelling microscope, like many scientific breakthroughs, depended on a happy confluence of talent, hard work, money and luck. In 1978 Heinrich Rohrer, a physicist at IBM Zurich Research Laboratory, hired a new research staff member, Gerd Binnig. Some weeks before his start at IBM, Binnig was listening to a lecture at a conference on low temperature physics in Grenoble, when his thoughts returned to an earlier discussion with Rohrer held while they were looking for a house in Zurich. Rohrer had raised the idea of studying thin oxide films on metals, but suggested that the appropriate tools were lacking. An old idea stirred at the back of Binnig's mind. Could vacuum tunnelling provide the key?

The idea of using vacuum tunnelling was not new. In 1972, Russel Young and co-workers at the National Bureau of Standards in Washington had outlined a device which had some similarity to what Binnig and Rohrer would ultimately build. Although this device, called a topografiner, usually had the tip much further from the surface so that the current was not a tunnelling current, they did attempt to obtain a tunnelling current. The two crucial aspects missing from the topografiner were the ability to scan the position of the tip and the very careful tooling of the tip used by Binnig and Rohrer.

When Binnig took the idea to Rohrer they soon realised that not only would it provide the basis for studying surfaces locally, but that if vacuum tunnelling was combined with scanning, they would have the basis for a radically new kind of microscope, a scanning tunnelling microscope that, in principle, would let them see individual atoms in a surface (see Figure 2.2). (It should be noted that the STM was not the first device to enable individual surface atoms to be imaged. This had been done much earlier by Mueller in 1956 using a field ion microscope.) In mid-January 1979, Binnig and Rohrer submitted their first patent disclosure on the STM.

The essence of the STM is, as the name suggests, quantum mechanical tunnelling. An electron in the surface of a metal can occasionally penetrate some distance beyond the classical turning point determined by the strong electrical forces in the surface. The chance of finding an electron in a classically forbidden region decreases rapidly with distance from the surface. In mathematical terms, the probability is an exponentially decreasing function of distance. If a conducting tip is brought up close to the surface of a metal, the probability of an electron tunnelling out of its electric prison in the surface and into the tip depends

Figure 2.2   Schematic representation of a scanning tunnelling microscope

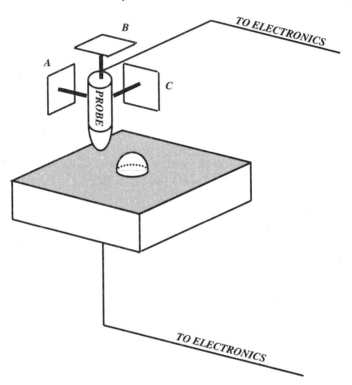

Note: The scanning tunnelling microsope (STM) uses the sensitivity of the tunnelling current to small changes in the separation between the probe tip and the sample. The tip can be moved in three perpendicular directions by small electrical devices, A, B, C. As the tip is scanned across the surface it encounters a bump, which reduces the separation between the tip and the probe. In order to keep the current constant the tip must be moved away from the surface. If this vertical movement is carefully monitored and correlated with the lateral position of the tip, a topographic map of the surface can be constructed.

very sensitively on the separation of the tip and the surface. A very small change in the separation could take you from a high probability of tunnelling to practically zero chance of tunnelling. Once the electron is in the probing tip, it can form part of a current. The more electrons that tunnel across into the tip, the larger the current. Thus the current in the tip will also depend very sensitively on the separation between it and the metal surface.

The surface of a metal is not perfectly flat and neither is the surface of the electron sea. In fact it is corrugated along the lines of the underlying atoms in the crystal. Electrons bunch up at an atomic core and thin out between the cores. If there is a small vertical step in the arrangement of the atoms in the crystal, the electron sea will also have a step. As we move the probe tip across the surface, the separation between the tip and the surface will vary, and thus the tunnel current in the tip will also vary. Because this current has such a sensitive dependence on the separation, the tunnel current will change dramatically as bumps and hollows in the surface of the electron sea are encountered (see Figure 2.2).

As we move the tip over the surface, suppose we try to keep the tunnel current constant. If we encounter a bump, we will have to move the tip away from the surface; if we encounter a hollow, we need to move the tip closer to the surface. By carefully monitoring the up and down motion of the tip we can map out the topography of the surface. It is a little like reading braille. However, it must be noted that the actual tunnelling probability depends not only on the separation of the tip and the surface but also on the density variations of the electron sea. This is a very important factor which complicates the interpretation of an STM image.

Needless to say, moving a tip up and down while it is moved across the surface is not an easy thing to do. For the first few months Binnig and Rohrer, working with talented craftsmen in their laboratory, struggled to overcome the difficulties and demonstrate vacuum tunnelling. It is easy to imagine some of the likely problems. How is one to overcome mechanical vibrations which would see the tip crashing uncontrollably into the surface? How strong are the forces between the tip and the sample and might these forces pull the tip into the electron sea? How to avoid changes in the length of the tip due to heating? Could the sample be moved on a very fine scale, up and down, over long distances? Finally, and perhaps most importantly, what should be the shape of the tip and how to make it? Given this list of potential problems, it is surprising that anyone would even try. Their first attempt was a complicated device which worked at low temperatures and ultra-high vacuum. It took a year to construct and to start testing the device, but in the end it was never used. It would take an additional seven years before the problems associated with a low-temperature ultra-high-vacuum STM were solved. Instead Binnig and Rohrer decided to make a simpler device which, in Binnig's words (Binnig and Rohrer 1987): 'used lots of scotch tape, and a primitive version of superconducting levitation wasting about 20 l [litres] of liquid helium per hour'.

As a theoretician I am often amazed at the ability of my experimental colleagues to go without sleep. It seems that the best experiments only work after midnight. I have often heard an experimental colleague preface his presentation at a conference with something like, 'we only managed to get this data on the night before I was due to leave for this meeting'. The first experiments on the STM were no exception. In the words of Binnig (Binnig and Rohrer 1987):

> Measuring at night and hardly daring to breath from excitement, but mainly to avoid vibrations, we obtained our first clear-cut exponential dependence of the tunnel current I on the tip-sample separation s characteristic for tunnelling. It was the portentous night of 16 March, 1981.

Binnig and Rohrer were excited, and Binnig set out to a conference in Los Angeles and a tour of some laboratories in the US. The response was good but not as enthusiastic as might have been expected for a development like this. It seemed that a device based on quantum mechanical tunnelling was just too exotic to be taken seriously. Meanwhile they turned their attention to an outstanding problem: the arrangement of atoms in the surface of silicon.

Deep inside a metal, or a semiconductor like silicon, the atomic cores are arranged in a regular array. Each atom is surrounded, on all sides, by other atoms, and electrical forces hold them in place. At the surface the situation is changed, and it should not be surprising to learn that the arrangement of the atoms can be very different to that in the bulk. The symmetry of the situation has altered; atoms no longer see the same world in all directions, and they rearrange themselves accordingly. This is called surface reconstruction and it depends on how the surface is aligned relative to the crystal structure of the bulk. The problem is to determine what the new arrangement is. The changed symmetry makes it particularly hard to calculate the new arrangement. The famous theoretical physicist Paul Dirac said, 'The surface was invented by the devil'.

You may think that the arrangement of atoms in the surface of silicon is a rather esoteric problem. In fact semiconductor technology depends on understanding surface reconstruction! To appreciate this you have to know how semiconductor devices are built. Typically, different semiconductor materials are deposited a layer at a time (a process known as epitaxial growth). The arrangement of the atoms in each layer, and the resulting electronic properties, depends on the arrangement of the atoms in the surface beneath, the substrate. Indeed, by carefully controlling factors such as temperature and impurity concentration, different surfaces can be formed to create different electrical properties in the layered structures (see chapter 4).

One famous example of surface reconstruction concerns the arrangement of atoms in a particular surface of silicon. Theory and some experiments, based mainly on electron microscopy, had indicated that the surface was a periodic

41

array of seven atoms by seven atoms. Different theories however predicted slightly different features for this reconstruction. Binnig and Rohrer realised that the STM was exactly the tool to resolve the issue. In 1982 they obtained a convincing image of this surface, and were overwhelmed by its beauty. As Binnig said, 'I could not stop looking at the images. It was like entering a new world. This appeared to me as the unsurpassable highlight of my scientific career and, therefore in a way, its end' (Binnig and Rohrer 1987). This is hardly an appropriate attitude for someone who has just commenced his career as a research physicist. However, anyone looking at these images can appreciate Binnig's feelings. (If you have access to the World Wide Web, you can view such images at the IBM Laboratory image library: http://www.almaden.ibm.com/vis/stm/lobby.html).

It was not long before the rest of the scientific community shared Binnig's elation. The paper on the silicon surface really convinced the community that the STM was an important breakthrough. So successful was this work that Binnig and Rohrer had difficulty finding time to continue their research, so inundated were they by requests for talks and visitors seeking to understand this new device. The STM, a device exploiting the strange quantum phenomenon of tunnelling, had arrived. In 1985 Binnig and Rohrer were awarded the Nobel Prize for Physics.

Atom-craft

In the past decade the STM, with its ability to see individual atoms, has revolutionised many fields, from the study of friction between surfaces in contact to biotechnology. Its application to the study of semiconductor surfaces is producing benefits in terms of better materials for new devices. The recently developed high-temperature STM has enabled the monitoring, in real time, of thermally activated chemical reactions, at temperatures up to 800^0C.

Very early on the device was used to try and image DNA, the biomolecular basis of life, triggering a rush of scientific publication. Here is a cautionary tale, the moral of which is, do not take a digitally regenerated image at face value. The basic problem is that DNA is not a conductor. To image DNA, it must be bonded to a special kind of surface, called a substrate. Fortunately DNA will stick to almost anything, or so it is generally believed. Early attempts to image DNA found that it was only possible to get good images if a special substrate called HOPG (highly ordered pyrolitic graphite) was used. This curious observation should have given some pause for consideration. Soon dozens of images of long chain-like objects meandering across surfaces, apparently containing some kind of periodicity, started to appear in (and on the covers of) international scientific journals.

In 1991, Thomas Beebe and Carol Clemmer of the University of Utah published images in the journal *Science*, which looked every bit like DNA chains. The only problem was that there was no DNA on the surface! This article carried a lot of weight as Thomas Beebe himself was one of the first to publish STM images of DNA. The lesson was clear. It was very possible that HOPG surfaces themselves could contain structures closely resembling chains of DNA.

These early problems have been largely overcome by the development of a new kind of scanning probe microscope, called the atomic force microscope (AFM), which does not depend on using an electrically conducting tip and surface. Instead a tiny cantilever is dragged across the surface. As it moves weak forces can bend it, and this bending can be detected by a variety of means. Today, scanning probe microscopy is a major tool for molecular biochemistry.

Perhaps the most important application of the STM is indicated in the work of another IBM laboratory. In 1990 Don Eigler and co-workers at the IBM Almaden Research Laboratories in San Jose used an STM to write the world's smallest word. In single atoms of xenon they spelt out 'IBM' on a nickel surface. The tip of an STM exerts a force on the atoms in the surface it is used to image. Eigler realised that by carefully adjusting the voltage on the tip and its separation from the surface, it would be possible to drag single atoms across a surface and place them almost at will. By exploiting the underlying periodicity of the nickel atoms in the surface, they were able to place xenon atoms on a rectangular grid that was four unit nickel cells long horizontally and five unit cells long vertically. This gave an image roughly one nanometre by one nanometre, which is a box with a side length of one-thousandth the width of a human hair. The possibility for atomic-scale computer circuits is immediately apparent.

Eigler's first demonstration of atomic calligraphy was astounding, but it suffered from a number of limitations. While arranging xenon atoms on a surface of nickel is impressive, it is unlikely to have much impact on a semiconductor technology based on silicon. In addition it required very low temperatures (four degrees above absolute zero, to be precise). Recently Mazakazu Aono and co-workers with the Atomcraft Project in Japan demonstrated manipulation of individual atoms on silicon surfaces (in fact the same 7×7 surface which figured in the development of the STM). Moreover, this was done at room temperature.

Unfortunately atomic graffiti is painstakingly slow. There is significant motivation, however, to increase the efficiency of the process. If one were able to store information at the level of, say, 1000 atoms per bit, the contents of an average library could be stored on a thumbnail-sized disk. The entire repository of human knowledge could be stored on a disk about 25 cm across.

The ability to manipulate atoms on surfaces will certainly lead to new technologies. We can get a glimpse of this in a recent experiment performed by Eigler and his group in San Jose. By arranging iron atoms in a circle on a surface of copper, Eigler formed a 'pond' for surface electrons. The diameter of the pond was about 7 nanometres across (that is of the order of 7 atoms across). To appreciate the significance of this we must recall

Planck's insight into what happens when quantum particles are confined. The copper electrons, boxed-up in the circular corral of iron, become like billiard balls in a circular billiard table. This confinement means that only certain energies will be allowed. If we were able to take a snapshot of the electrons in the corral we would see a two-dimensional analogue of the one-dimensional billiard ball in chapter 1. There would be periodic variations in the density of electrons across the surface of the corral, rising from a minimum at the boundaries, through a sequence of peaks and troughs to a maximum in the centre. But a snapshot of electron density is exactly what the STM can produce when it is scanned across the surface! Using the STM, Eigler was not only able to place the iron atoms in a circle on the surface, but able to image the electrons trapped inside the corral. The image provides astonishing confirmation of the predictions of quantum mechanics. In addition, by using the ability of the STM to probe the energies of electron surface states, Eigler was able to demonstrate that the energies of the trapped electrons were restricted to a discrete set of values.

The STM has become our hands and eyes to explore the quantum world. Imagine the experience if an STM were linked to a virtual reality headset and joystick. If some feedback could be arranged between the forces encountered by the tip on the surface and the joystick, we could literally feel and see individual atoms. This may sound a little far-fetched, but in fact such a device, called the nanomanipulator, already exists at the University of North Carolina at Chapel Hill (UNC–CH). The nanomanipulator was conceived by Warren Robinett then at UNC-CH and Stan Williams then at University of California. Using virtual reality technology, including a stereoscopic colour mounted headset, and force feedback remote manipulator, it creates the illusion of a surface floating in space before the user. Hand movements are converted to computer commands and sent to the STM, while a returning signal allows the user to see and feel the surface topography.

45

Our understanding of the every day world is built up from manifold experiences mediated by our senses. It is often said that quantum mechanics looks so difficult because we cannot have any direct sensual contact with the atomic realm. The nanomanipulator indicates that this may not always be so. Perhaps we can see and touch the quantum world. Like a newborn child, our interior image of this world would be built up through action and play. Sitting by a sea of electrons, we could pile atoms on top of atoms in atom-castles. We could cast atoms into the electron sea and watch the ripples move across the surface. We could almost live the strange world of the quantum.

CHAPTER THREE

Prisoners of Light

In the basement of my department at the University of Queensland are held captive a small collection of rubidium atoms, trapped in an ultra-cold prison of light and forced to give up the secrets of the quantum world. Using subtle forces produced by laser light, atoms are pushed and pulled, held in a tight embrace and brought to a virtual standstill. Frozen to temperatures barely one-millionth of a degree above absolute zero, the atoms move according to the rules of the quantum world. In this state, they no longer behave like little billiard balls, but exhibit instead the constrained lawlessness of quantum uncertainty. Atoms bend around laser beams like waves breaking on a rocky outcrop, and bounce vertically from surfaces of pure light. Single atoms, manipulated by beams of light, follow an intricate chaotic dance, modulated by the quantum rules underlying the classical realm. Similar atom traps may now be found in many laser laboratories around the world. Barely ten years old, this new technology forms the basis of one of the fastest growing areas of modern physics—atom optics (Chu 1992.)

How low can we go?

The existence of atoms as fundamental constituents of matter, while postulated in ancient times, finally became orthodoxy only in the early stages of the twentieth century.

Of course by the late nineteenth century most physicists had accepted the concept, but not all. Through the work of Boltzmann, Maxwell and Gibbs, the atomic hypothesis showed how thermodynamics, the great triumph of nineteenth century physics, could be reconciled with the physics of Newton, supplemented where necessary by statistical laws. Heat was explained as the incessant random motion of large numbers of atoms. Removing heat from a body results in the constituent atoms moving more slowly on average. The temperature of an object is a measure of the average energy of motion of its atomic or molecular subsystems.

What is the lowest temperature we can reach? According to classical mechanics it is always possible to bring an atom to a total stop. This is not easy to do, especially for the myriad of atoms that compose a macroscopic sample of matter. But if it could be done the temperature must, by definition, be zero, or 'absolute zero' to give it the appropriate technical designation. By extrapolating a common temperature scale, such as the Celsius scale, we can assign a numerical value to absolute zero. In Celsius it is about −273°C. Throughout this century an entire branch of physics has developed dedicated to achieving lower and lower temperatures, the reason being that slow atoms are much easier to study and control. A large number of interesting phenomena, such as superconductivity, are revealed at sufficiently low temperatures.

Quantum mechanics requires some modifications to the idea of absolute zero. To bring an atom to rest we must apply constraining forces which tend to localise it in some region of space. In the case of the atoms that make up a crystal, interatomic forces hold the atoms in a strict array. The atoms are thus forced to occupy a small region of space. As we saw in chapter 1, any attempt to confine a particle in space restricts also the allowed energies it can have and particularly implies that there is a lowest energy. Any attempt to remove energy beyond the lower limit cannot succeed. Thus quantum mechanics makes it impossible to slow an atom completely to rest. The eternal, unavoidable fluctuations in the position of a particle, no matter how tightly confined, cannot be stopped. Thus in a quantum world we can never cool a particle to such an extent that it is completely frozen in space. The best we can do is to reduce all particles to the lowest energy state allowed by quantum mechanics, called the ground state.

Many ingenious schemes have been proposed to reach temperatures approaching absolute zero. Of these perhaps the most surprising is the idea that beams of light can be used to cool atomic gases. We normally associate light, especially sunlight, with warmth. In popular fiction and indeed reality, laser beams are seen to burn through all manner of substances. In atom optics, however, laser light is used to chill material to near absolute zero. To understand how this can be we need to consider the forces that light exerts on matter. As you will see, it is a complicated business. Don't expect to be able to buy a laser wine cooler in the near future.

The force of light

In his explanation of the photoelectric effect Einstein showed that in some cases we must regard the energy carried by light as transported by little particle-like packets called photons. In the photoelectric effect an individual photon of the right colour may carry enough energy to liberate an electron from the ions in the surface of a metal. The electrons in atoms of a gas also can interact with individual photons of light. It was the study of how atoms absorbed and emitted light that enabled Niels Bohr to lay the foundations for quantum theory. The crucial observation is that atoms cannot emit or absorb light of just any colour. Rather the frequencies are determined by the energy differences between the allowed energies of the confined electrons surrounding the atomic nucleus. If a photon has just enough energy to match an energy difference between two allowed electron energies, the photon will be absorbed by the atom.

The energy of the electron is changed in the process of absorbing a photon and so is the energy of the entire atom. From the perspective of the atom, it is as if a small energetic particle, the photon, has been caught and held. Not surprisingly the atom reacts by recoiling just a bit. Likewise if an atom emits a photon (when an electron falls from a higher allowed energy to a lower) the entire atom also recoils. Combining Newtonian physics with the photon picture of light we are led to the conclusion that absorption and emission of light can have a mechanical effect on the entire atom (see Figure 3.1).

In fact, the motion of an atom plays a big role in determining how it interacts with light. As Einstein explained, the energy of a photon is determined by its frequency. But the frequency seen by the atom depends on how fast and in which direction it is moving. This is the optical analog of a familiar effect, the Doppler effect. It is a common observation that the perceived pitch of a sound depends on the relative motion of the source and receiver. A police siren falls in pitch just as the police car passes us. A similar effect occurs for light. The change in the frequency of light for moving sources provides us with the evidence that the universe is expanding. Light from distant stars is shifted towards the red, that is to lower frequencies, as those stars are moving away from us at an enormous velocity. If a star was moving towards us, the light would be shifted to higher frequencies, that is towards the blue. If the stars in the heavens were observed to have a blue shift we would have cause for concern.

A moving atom thus sees a different frequency to one standing still. In a gas the atoms have a range of velocities distributed around an average velocity which determines the temperature. Thus not all atoms see the same frequency of light. Some atoms will see photons of just the right frequency and absorb them, while other atoms, with a

Figure 3.1   Laser cooling of atoms

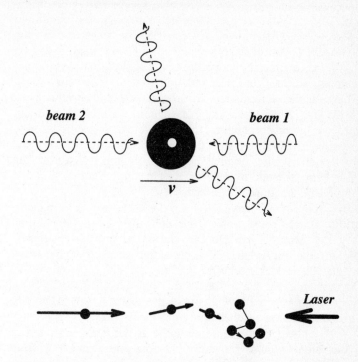

Note: An atom moving into a laser light beam sees a shorter wavelength (higher frequency). If the velocity is just right, the shifted frequency can cause an electron in the atom to absorb a photon from the light (top figure). As a result the atom recoils in the opposite direction. The excited electron eventually relaxes back to its original state, emitting a photon in the process. However, this spontaneous emitted photon can be scattered in an arbitrary direction, and so on average does not change the average speed of the atom. The net effect is to slow the atoms to a crawl after which they execute a random walk due to spontaneous emitted photons (bottom figure).

different velocity, will see the same light source shifted in frequency and thus will be unable to absorb a photon.

Once an atom has absorbed a photon it cannot hold it forever. Once absorbed, the photon vanishes and the electron in the atom is promoted to a state of higher energy. But the quantum theory of light predicts that the higher energy state is unstable and that the electron must decay back to the ground state, a process known as spontaneous emission. As it decays the atom emits a photon with an energy corresponding to the change in electronic energy, but in an arbitrary direction, not necessarily in the same direction as the original absorbed photon. Whatever the direction the atom chooses to emit the photon, it must recoil in the opposite direction.

Suppose we have an atom in a gas which is moving into a laser beam at just the right speed for the photons in the beam to be Doppler-shifted to match an internal electronic energy gap. If the photon is absorbed the atom gets a little kick backward and the electron is promoted to a higher level. The result of the little recoil kick is to slow the atom just a bit. Of course the electron will eventually decay, emitting another photon in the process. If this photon were emitted in the same direction as the incoming photon, the atom would recoil in the opposite direction, thus undoing the small slow-down it experienced by absorbing the photon. But a spontaneously emitted photon can be emitted in an arbitrary direction, and in fact is very unlikely to be emitted in the same direction as the photon it absorbed. The atom recoils, but in a different direction, and does not undo the slow-down in its motion towards the laser beam. On average the net change in velocity due to spontaneous emission is zero. The net effect of many such absorptions and re-emissions would be to slow the atom in its motion towards the laser beam, while adding a small random motion as it emitted photons spontaneously in other directions. Looking at this from the outside it is as if the atom ran into a hail of bullets, which it swallows and regurgitates in random directions; the net effect of this scattering of photons is to slow the atoms moving against the laser beam.

The scattering force is quite small. The change in velocity due to a single photon absorption is about one centimetre per second. This should be compared to the speed of an atom in a gas at room temperature, which is about the same as a supersonic jet. The scattering force was first observed by Otto R. Frisch in 1933, by deflection of a beam of sodium atoms. The beam was produced by vaporising sodium in a small oven and allowing atoms to pass through a series of small slits after the opening. The light used was generated by a sodium incandescent lamp. Such a weak source has only a tiny effect on the atoms as flux of photons is too low to give many absorption and scattering events. A much larger effect can be achieved by using laser light. An early observation of laser deflection of a sodium beam was made by Wang Yu-Zhu and his collaborators at the Shanghai Institute for Optics and Fine Mechanics in 1984.

The scattering force (it is sometimes called the spontaneous force because of the role played by spontaneous emission) is not the only force that light exerts on matter. Another force called the dipole force can also be used to control the motion of an atom. In fact we are all familiar with dipole forces. When a plastic rod is rubbed on some material it acquires an electric charge. Such a piece of plastic can exert a force on small objects such as pieces of paper. What happens is that the electric charge on the plastic causes charge to separate on the otherwise neutral piece of paper. The charge on the paper nearest to the rod is opposite to that on the rod, and thus the paper is attracted.

An atom is composed of an outer shell of negatively charged electrons and an inner core of positive charges. Left alone the atom arranges itself so that the positive and negative charges balance over space to give an electrically neutral particle. Since it is neutral, it will experience no electrical force if placed in an electric field. However, if we apply a strong electric force to the atom, say by bringing up a large store of negative charge, we will distort the carefully balanced arrangement of charge. Some of the electrons may be repelled by the electric force, leaving a small net positive charge on one side of the atom. What results is called an induced electric dipole as it has a small negative charge on one end and an equal positive charge at the other end. We say that the atom has been polarised. Though overall electrically neutral, the unequal spatial distribution of charge on the atom means that it will now feel a force due to the applied electric field if the field is not spatially uniform. The net effect is that the atom is attracted to regions of strong electric field. This dipole force is just the kind of force needed to trap an atom by light.

In the late nineteenth century James Clerk Maxwell showed that light was a self-sustained oscillation of electric and magnetic fields. Light itself can then induce an electric dipole in a neutral atom.

The oscillating electric dipole will follow the oscillations of the light so long as the frequency of the light is below that of the natural frequency of the atomic dipole. The natural frequency of the dipole is proportional to the energy gap between the transitions excited by the laser light. If the frequency of the light is greater than the natural frequency of the dipole, the dipole will be totally 'out of phase' with the light, that is to say the oscillations in the charge separation lag behind the changes in the changing electric field to such an extent that the charge distributions are the reverse of the low frequency case. This means that the force on the dipole will also reverse as the frequency of the laser is increased above the natural frequency of the dipole. If the light has a frequency lower ('red' detuning) than that required to match an allowed energy gap in the atom, the force tends to push the atom to regions of higher intensity. If the laser is tuned towards the 'blue', that is higher than that required to match an allowed energy difference, the force will push the atom towards regions of lower intensity. This dependence on laser frequency means great control can be exercised by using the dipole force.

One other thing is required: the atoms must be moving sufficiently slowly to be trapped by the induced dipole force. This is where laser cooling enters the picture. The first suggestion that optically induced dipole forces could be used to move and indeed trap atoms was made by the Russian physicist Vladilen S. Letokov in 1968 and in 1978 Arthur Ashkin at AT&T Bell Laboratories gave a more practical scheme for trapping using focused laser beams.

Light up and chill out

Doppler shifts are the key to using the spontaneous light force to slow atoms. If we use a beam tuned just below an atomic energy gap, an atom moving into the beam will see its frequency increased just enough to match an allowed energy gap. This atom absorbs a photon from the beam and is thus slowed down. The scattering force on the atom thus depends on its velocity and it is this which makes it possible to cool fast atoms with laser light. For a wide class of atoms, the spontaneous scattering force is sufficient to bring atoms emitted from a hot thermal source to almost a standstill in a fraction of a second. In 1975, D. Wineland and H. Dehmelt first suggested that a velocity-dependent scattering force could be used to cool atoms. A similar idea was proposed by Ted Hansch and Arthur Schalow at about the same time. The first observation of how the scattering force could be used to cool atoms with lasers was made by two separate groups led by Bill Phillips and Jan Hall at the National Bureau of Standards in 1985. In this case the photon scattering rate was increased by 10 million compared to the early work of Frisch. The result is a deceleration corresponding to 100 000 times the acceleration due to gravity.

In a gas atoms are moving in all directions, and a single laser beam will not be sufficient to cool all the atoms. An atom moving away from the laser beam will see a frequency that is Doppler-shifted away from an internal electronic energy gap. Such an atom cannot be slowed, as it will never absorb a photon. To fix this we can take another laser beam from the opposite direction. Clearly we needn't stop there. Better still is to use six laser beams in three oppositely directed pairs along the three spatial directions, north–south, east–west and up–down. The resulting arrangement is known as the optical molasses, as fast atoms entering such a configuration of lasers will suddenly find themselves moving in a kind of thick viscous fluid. As atoms are slowed they will of course see a different Doppler shift. Thus some arrangement must be made to keep the laser frequency just right for a moving atom to see a photon of just the right energy to absorb. The first demonstration of the cooling effect of an optical molasses was made by Arthur Ashkin, Leo Hollberg, John E. Bjorkholm, Alex Cable and Steve Chu at AT&T Bell Labs in 1985. They were able to cool sodium atoms to 240-millionths of a degree above absolute zero.

It was well understood that an optical molasses could not cool atoms an arbitrary amount, due to the random recoil when photons were spontaneously emitted. The lowest temperature expected, the so-called Doppler limit, is about 240-millionths of a degree above absolute zero for sodium. However, some of the first experiments conducted by William Phillips and his colleagues at the National Bureau of Standards in Maryland produced a big surprise. The actual temperatures reached were much lower than the theory predicted. This astonishing fact produced a flurry of experimental and theoretical work, first to explain why the scheme worked so well, but, more importantly, to exploit the apparent availability of extremely cold atoms. Cold atoms are slow atoms and thus subject to the quantum rules. The success of laser cooling takes us a step closer to a technology based on the quantum motion of entire atoms.

The theoreticians were quick to catch up with the experiments. Jean Dalibard and Claude Cohen-Tannoudji of the Collège de France and the École Normale produced an elegant theory which explained sub-Doppler cooling. A similar theory was produced by P. Ungar and co-workers in Steven Chu's group at Stanford. The essential component of these theories was the realisation that multiple electronic energy levels can participate in the absorption and emission of light. This work showed that it would be possible to cool atoms to a velocity equal to the recoil velocity—the velocity imparted to an atom due to the emission of a single photon.

I will explain one form of sub-Doppler cooling known as the Sisyphus effect (Cohen-Tannoudji and Phillips 1990). Sisyphus was the unfortunate victim of the ever-truculent Greek gods, condemned to spend his life rolling a large stone to the top of a hill, letting it roll back down and starting all over again. Needless to say one would soon become rather short of energy in this activity. In Sisyphus cooling the atom moves in the standing wave formed by two laser fields propagating in opposite directions. The intensity of the light in a standing wave has a wavelike spatial variation. In some places, called the nodes, the intensity is zero while in other places, the antinodes, the intensity is a maximum. An atom sitting at a node experiences no dipole force. If the atom was not sitting exactly at a node, however, it would feel a dipole force pushing it to regions of lower intensity, that is back towards the node. If we take an atom at a node and displace it slightly it will roll backwards and forwards around the node. The atom behaves like a ball rolling at the bottom of a very smooth valley. If we put the atom at an antinode and then give it a little push, it will start rolling across all the peaks and troughs (see Figure 3.2), speeding up as it rolls down the side of the standing wave, pushed by the dipole force, only to slow down again as it moves up the next peak. On average, the atom does not speed up or slow down, so this scheme cannot be used to cool fast atoms. Somehow we need to arrange for the atom to be always moving uphill even as it moves over large distances in the standing wave.

Figure 3.2　Sisyphus cooling

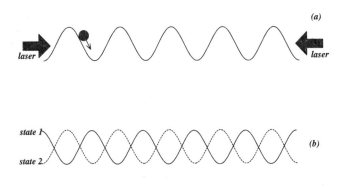

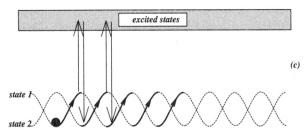

Note: Two oppositely directed lasers set up an intensity pattern which varies periodically in space. An atom moving in such a beam feels a dipole force which periodically opposes the motion and then enhances the motion. Effectively the atom is like a small ball rolling over a sequence of identical hills (a). The dipole force on the atom depends on which state it is in, and is equal and opposite for the two states. This means one state sees a valley where the other state sees a hill (b). By carefully switching the atom from one state to the other (using higher excited states) we can arrange for the atom to be always moving uphill, just like Sisyphus in the Greek myth (c). The atom thus gradually slows down as its initial energy is carried away by the scattered light.

The way to do this is to suddenly transfer the atom back to a valley every time it gets to a peak. We need not worry about the details. The basic idea is to use laser light and magnetic forces to rapidly rearrange the electrons inside the atom so that their electric dipole is suddenly reversed. If this rapid rearrangement is always done when the atom is at a peak, instead of seeing a downhill run on the other side it again sees a hill to climb. The atom is condemned to be always rolling uphill until eventually it slows to a stop. I think you will agree that Sisyphus cooling is a rather good name for this process. Further details are given in Figure 3.2.

For cesium atoms, sub-Doppler cooling meant temperatures of about three-millionths of a degree above absolute zero could be reached. Since the discovery of sub-Doppler cooling many novel suggestions have been made to cool below the recoil limit, and some implemented. The record for laser cooling changes a couple of times a year. In mid–1995 it was held by a group at the University of Colorado, with a temperature of about 0.00000003 of a degree above absolute zero.

Atom traps

The optical molasses and more sophisticated schemes are able to produce small, very cold bunches of atoms, but they cannot hold the atoms in space indefinitely. Because these cooling schemes rely on the scattering force, which is always proportional to the intensity, sooner or later an atom will be forced out of the cooling region. There are two ways to avoid this problem. One way is to modify the scattering force so that it is not simply proportional to the intensity of the lasers. Another way is to use the dipole force. To trap an atom we need the optical equivalent of a particle on the end of a spring. Any attempt to move the particle extends or compresses the spring. This results in a force which tries to restore the particle to its original position. A force of this kind is called a restoring force. The optical dipole force is just such a force.

If the dipole force were the only force an atom in laser field experienced, trapping would be a fairly simple matter. However, as we know, there is also the scattering force due to absorption and emission of photons. This is the force responsible for laser cooling. Unfortunately the two forces always come together. We could try to minimise the scattering force by making sure the frequency of the laser light did not match very well an allowed energy difference for the electrons. This reduces the chance that the atom will absorb a photon. Not surprisingly it also reduces the size of the dipole force. The trapping force becomes so weak that even laser-cooled atoms could not be trapped. Furthermore the dipole force is quite weak and can only trap quite slow atoms. To make use of the dipole force we need to slow (that is cool) the atoms. The crucial innovation was the development of the magneto-optical trap (MOT).

The first successful experiment to trap neutral atoms made use of a well-known fact—some atoms behave like a small magnet. If we place such a little magnet in a magnetic field, which varies in space, the north pole of the magnet will experience a different force to the south pole, leaving a net force which can move the atom. Using a careful arrangement of magnetic fields, and laser cooling techniques, the NIST group led by Phillips built the first neutral atom trap in 1985. It was not long before the idea of combining magnetic fields and optical forces occurred to researchers. The first hint on how this could be done was given by David E. Pritchard at Massachusetts Institute of Technology and Carl E. Wiemann at the University of Colorado. They indicated that a spatially varying magnetic field could be used to significantly modify the scattering force due to absorption of laser light. Shortly thereafter Jean Dalibard of École Normale Supérieure in Paris proposed a design for a magneto-optical trap using multilevel atoms. The first trap of this kind was demonstrated in 1987 by a collaboration between Pritchard's group at MIT and Chu's group at AT&T. This kind of trap now occurs as an essential component in most atom optics laboratories. In 1994, for example, it was estimated that there were over 150 MOTs in operation.

The idea behind an MOT is simple. The electrons in an atom can oscillate at many frequencies and can thus 'tune-in' light of many different colours. Usually we vary the frequency of the light until we hit just the right one for the electrons to respond. In an MOT we do it the other way around—we tune the electrons in the atoms directly, while keeping the frequency of the light fixed. To change the frequencies at which the electrons will respond we use a magnetic field. Increasing the strength of the magnets can increase or decrease this frequency in a precise and controllable way.

Now suppose we use two beams of light with different frequencies and shine them on the atoms from opposite directions. If the electrons can respond to only one frequency at a time they will absorb photons only from the beam that has the right frequency. If they do absorb light from only one beam they experience a scattering force which pushes them along in the direction of that beam. The trick is now to 'tune' the electrons in the atom so that first they absorb light from one beam and get pushed in that direction, and then they absorb light from the opposite beam and get pushed in the other direction. To get a picture of this, imagine if this sort of thing happened to your radio as you tuned in different stations located in different parts of the city. If you tuned in a radio station to the north your radio would be pushed to the south; if you tuned in a radio station from the south your radio gets pushed to the north. As you fiddled with the tuning knob, your radio would be sliding around all over the place. Fortunately this does not happen for heavy objects like radios, but it can and does happen for atoms as they tune in light of differing colours.

To tune the atoms we arrange for the magnetic field to vary along the line the atoms are moving. As the atom moves it sees a different magnetic strength, At one point it is tuned to absorb light from, say, the left and gets pushed to the right. As it moves to the right it eventually reaches a region where it sees a different magnetic field, tunes in the light from the other oppositely directed beam, and gets pushed to the left. The atom oscillates backwards and forwards, pushed by one light beam then the other. The result is that the atom tends to oscillate backwards and forwards between the two points where it can absorb light. Further details are given in Figure 3.3.

Because the trapping in an MOT is based on the spontaneous force it still has the characteristic dependence

Figure 3.3   Principles of a magneto-optical trap

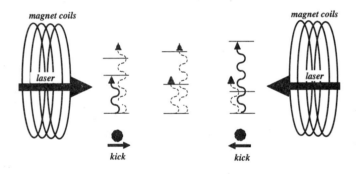

Note: A magneto-optical trap (MOT) forces an atom to oscillate back and forth between two points where it can absorb a photon, first from one beam and then from the other, oppositely directed, beam. A magnetic field is used to change the conditions for an atom to absorb a photon. At the left turning point the energy of the left-going photon matches the energy difference of an electronic transition. This photon is absorbed and the atom recoils to the right. At the right turning point the magnetic field adjusts the electronic energies so that another set of electron energies can only absorb a photon from the left-travelling beam. The atom then recoils to the left.

on velocity required for cooling. This means the MOT not only traps but cools atoms simultaneously. It also has a large region of space from which it can capture moving atoms. This means a substantial density of trapped atoms can be achieved. An MOT for cesium can be filled with one thousand million atoms in less than a second from room temperature vapour. Typically the velocity of these atoms is of the order of 2 centimetres per second. The MOT provides an ideal source of slow atoms. Just why slow atoms are interesting will be discussed below.

It is worth mentioning here a related technology currently emerging in a number of laboratories which also makes use of the scattering and dipole optical forces, not to trap and control the motion of atoms but rather to control larger particles such as living cells. These technologies generally go by the name of optical tweezers. The forces light exerts on these particles are essentially the same as those that can be exerted on an atom. The ability to manipulate small particles such as cells has some interesting applications. Many of these applications were first suggested almost twenty years ago by Arthur Ashkin, one of the pioneers of the field of optical forces. In the laser micromanipulation laboratory at the University of Queensland we are using optical tweezers to separate different species of bacteria suspended in a liquid—without killing them in the process. Another possible application we are pursuing is to carefully manipulate brain neurons to construct purpose-made neuronal circuits with a specific function. Such function-specific circuits may prove to be a useful means to test the effects of new drugs. Recently our laboratory has shown that the scattering force can be used to rotate small particles. This is done by giving the light beam a 'twist' by passing it through a carefully constructed holographic plate. This may be a means to power small micro-machines by non-mechanical means, or perhaps to produce micro-stirrers for micron scale chemistry.

The serendipitous discovery of sub-Doppler cooling combined with trapping techniques enables small clouds of very cold atoms to be suspended in space. I am always surprised when I see these little collections of cold atoms (about one thousand million) fluorescing happily in their cold prison of light. That something as tiny and insubstantial as an atom can be held at will by an ingenious technology is a neverending source of amazement to me. But these devices were not built to amuse credulous theoreticians such as me. In fact, the ability to control the motion of single atoms has opened up an extraordinary new field of physics known as atom optics.

Atom optics

The technological innovation of laser cooling and trapping has given us a means to force atoms to abandon the ways of classical physics and adhere to the rules of the quantum. For most people, from antiquity to the present day, an atom is something like a little billiard ball, and behaves in pretty much the same way. We now know of course that it has a complex internal structure with electrons confined in the electric grip of a positive nucleus. Yet despite this internal structure, most of the time we will get into little trouble if we persist in imagining the atom as a whole to behave like a classical particle. Ultimately our idea of a billiard-ball atom stems from many experiments in which atoms do indeed behave like localised particles with a definite velocity. In the physics of Newton, the physics of the everyday world, it is assumed that measurements can be made arbitrarily accurate in principle even if in practice they fall somewhat short of this ideal. Yet quantum theory indicates that this can only be an approximation. In fact all measurements to determine the position and velocity of an atom will demonstrate an irreducible randomness, not due to an inability to perform accurate measurements, but rather due to the intrinsic uncertainty of the quantum world.

In Newtonian dynamics if we know the position and velocity of an atom at any instant and we know the forces acting on it we can predict where it will be and how fast it will be moving at any time in the future. The theory does not refer to the outcome of measurements explicitly precisely because it assumes that any measurement can be as accurate as we want. We now know that this is not the case. There is an irreducible randomness in determining the position and momentum of any particle. Even if we carefully prepare identical experiments every time, quantum randomness will result in the 'scattering' of measurement results around some mean value.

In the early part of this century, after many puzzling experiments had been carefully analysed, the quantum theory was invented to describe in detail the nature of this irreducible randomness. To do this the theory had to present a definite mathematical scheme to calculate the probabilities, the 'odds', for the various results of position or momentum experiments. It also had to explain how these probabilities would change when forces acted on the particles in question. So unlike Newtonian mechanics, which is expressed in terms of where a particle is and how fast it is moving under the action of various forces, the quantum theory is expressed entirely in terms of the probability for a particle to be found at a particular position with a particular momentum under the action of the same forces.

So far as we know, quantum theory applies to all particles no matter how large or small they happen to be. Yet clearly, for many everyday macroscopic observations, Newtonian mechanics suffices, and we don't need to worry about the intrinsic uncertainty in position and momentum measurements. The reason is that, compared to the typical distances and velocities involved in experiments on macroscopic systems, errors due to quantum uncertainties are really very, very small. The measuring devices we build are designed to detect changes in position and velocity which are much larger than quantum uncertainties. There is no practical need to build them more accurately. The sizes of the objects themselves are typically vastly larger than any quantum uncertainty in position. To access true quantum behaviour we need to arrange for the changes in position

Figure 3.4   An atom interferometer

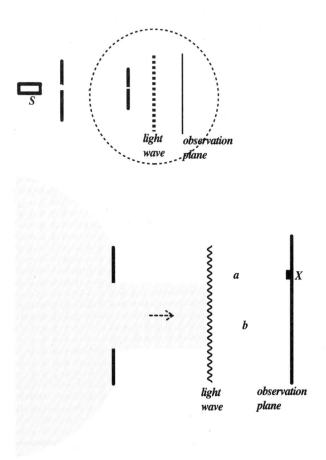

Note: In an atom interferometer a source of atoms, S, is collimated by passing the beam through a sequence of slits to produce a broad beam of atoms all heading in roughly the same direction. The atoms now encounter a deflecting grid, which could simply be a spatially varying light beam. The width of the atomic beam is much wider than the spatial variation of the deflecting grid. Finally an atom detector is placed at a distant screen to count the number of atoms arriving. As an atom arriving at a given point, X, could have travelled many paths of different lengths, we must first add the probability amplitudes for each path considered separately. This can lead to a great deal of cancellation between the constituent amplitudes. Only for certain special directions do the amplitudes add together to give a significant detection probability. In contrast, an atom behaving like a classical billiard ball could arrive at any point on the screen.

---

and momentum to be of the order of the intrinsic quantum uncertainties in the system. In practice this means very small objects moving very slowly and it is here that laser cooling and trapping of atoms has a role to play.

The problem of producing slow atoms has largely been solved by the discovery of sub-Doppler cooling and the availability of atom traps such as the MOT. However, in order to do anything interesting we must be able to apply forces to these same atoms. In other words we need to 'get a handle' on cold trapped atoms. The key is the optical dipole force. In atom optics the particles are pushed around and manipulated by light itself.

In the case of the dipole force, a laser beam distorts the electron distribution surrounding the atom to produce a small charge separation, called a dipole. Once formed the dipole can itself interact with the strong electric field in the laser light. The crucial result is this: for light at a frequency lower than a nearby electronic energy gap, the atom will be attracted toward regions of greater intensity; for light at a higher frequency than a nearby energy gap, the atom will be repelled from regions of higher intensity. It is very easy to produce light with an intensity variation over a small region of space. For example, two counterpropagating light waves will produce a standing wave in which the intensity oscillates from a zero to non-zero values on the order of half an optical wavelength, a distance of the order of one-tenth of a millionth of a metre. The atom will thus experience a force which has a rapid variation in space.

67

The ease with which light can be modulated in space gives us a great deal of flexibility in constructing interesting arrangements of dipole forces to deflect atoms.

One of the first experiments in atom optics was the observation of atom diffraction by an optical standing wave. In 1983 Phillip E. Moskowitz and co-workers in David Pritchard's group at MIT made the first observations of the diffraction of sodium atoms. This kind of experiment does not require particularly slow atoms, in fact slow atoms would complicate the picture considerably. But it does require atoms with a well-defined velocity. In the MIT experiment this was done by using slits to spatially collimate a beam of sodium atoms carried along in a supersonic beam of argon. The basic configuration is illustrated in Figure 3.4. The dipole force in the standing wave deflects atoms by pushing them away from regions of high intensity. The atoms move sufficiently fast through the standing wave that it is unlikely that an atom will oscillate backwards and forwards between regions of high and low intensity. (This is why cold atoms are not really necessary.) They simply experience a 'kick' as they pass through the beam. An atom arriving at a particular point on the distant screen could have come from a number of different places, and thus may travel different distances to the screen, as in Figure 3.4.

To get the total probability of detecting an atom at some point we need to compute the probability amplitudes for each way considered separately and add them together as little arrows (see chapter 1). All the arrows undergo a different rotation, as the length of each path is different. In some cases most of the arrows reinforce each other to give a large probability for detection. At other points on the screen the arrows tend to cancel, giving low probability for detection. The result is that the patch of atoms falling on the distant screen contains dark bands where no atoms are ever detected. The MIT group saw clear evidence for these bands. A more refined experiment, using helium, was done by the atom optics group at the University of

Konstanz in 1993 and again clear evidence for quantum probability amplitudes was observed.

Atoms can also be reflected by optical fields. An experiment based on this idea was suggested by Geoff Opat of the University of Melbourne in 1989. When light passes from glass into air or vacuum it is partially transmitted and partially reflected. If the angle of incidence is carefully adjusted it is possible for it to be totally reflected, a phenomenon know as total internal reflection. The intensity of the light outside the glass does not go immediately to zero, however, but falls off exponentially over the order of a wavelength of light. Such a light field is called an evanescent wave. A slow atom encountering the evanescent wave will see an exponentially rising force pushing it away from the surface. If the incoming velocity of the atom is low enough it can be completely reflected by the evanescent wave, thus forming a kind of atom mirror.

In 1982 R. J. Cook and R. K. Hill, then at the Air Force Institute of Technology in Ohio, were the first to suggest that an atom mirror could be made with evanescent light and the first observation was made by V. I. Balykin and co-workers in the former Soviet Academy of Sciences in 1982. If we simply drop cold atoms vertically onto this 'light-mirror' they can bounce up and down like a child on a trampoline. The basic idea is to drop cold atoms from the atom trap (by turning off the confining beams) onto a glass surface supporting an evanescent wave (see Figure 3.5). Atoms reflected from the surface are returned to the surface by the action of gravity and can then perform a series of bounces. We have a situation where the atoms are confined to some region above the trampoline and, like all quantum confined objects, will be subject to the quantum speed restrictions. In practice this means that bouncing atoms can only have particular allowed energies. In 1993 a group at École Normale Supérieure in Paris were able to observe as many as ten bounces before atoms fell off the trampoline. However in this experiment the atoms were released at a comparatively large distance from the surface.

Figure 3.5   A trampoline for cold atoms

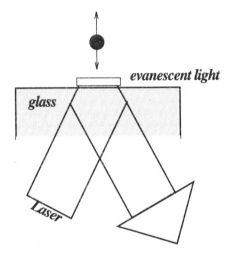

Note: Light reflected at a glass–air interface penetrates a small distance into the air before being totally extinguished. This decaying light beam can be used to deflect atoms falling onto the surface. The result is a small trampoline for atoms.

This means that the spatial extent of the bouncing atoms is so large that the quantum speed restrictions are not very restrictive and many possible energies are allowed. To see the restriction of energies predicted for quantum confined objects, the atoms would need to be released much closer to the surface.

If the atom were released close to the surface and many bounces could be achieved, a very interesting effect predicted by my atom optics group at the University of Queensland could be seen. Suppose we start with a ball of atoms initially confined to a very small region above the surface. Different atoms in the cloud will be at different heights above the surface. Atoms falling from different heights take different times to return to their initial position.

Classical mechanics predicts that, after some time, the ball of atoms will become spread into a persistent cloud, stretched between the surface and the release height. Our quantum calculations show, however, that the cloud will not persist. After a certain time the smeared-out cloud will gather itself back into a little ball almost identical to the initial distribution. An observation of this effect would be a striking demonstration of how different classical and quantum dynamics can be.

Another very interesting way of using evanescent light to control the motion of atoms is being investigated by Craig Savage at the Australian National University, in collaboration with a group at the University of Colorado. Savage proposes using a hollow optical fibre to support the evanescent light. The idea was first suggested by M. A. Ol'Shanii and co-workers at the Russian Academy of Sciences in Moscow in 1992. The hollow fibre is essentially a tiny glass tube. Solid optical fibres are now routinely used for broadband optical communication systems. Light can be launched down the glass cylinder forming the walls of the cylinder. An evanescent light field then coats the inside of the hollow fibre. Atoms sent down the fibre will be repelled from the walls of the fibre by the light forming a small pipe, or hose, for cold atoms. As the fibre is very flexible, in principle the atoms can be piped around the laboratory.

Another way to do it is to use light at a frequency *below* the electronic frequency, referred to as red-detuning. In that case the dipole force tends to attract atoms to regions of higher intensity. We can then send a laser beam, with red-detuning, down the centre of the fibre and, due to reflections from the walls, it will follow the curves and twists of the fibre as it propagates. Atoms in the hollow core of the fibre will tend to stick to the centre where the laser light has a high intensity and thus also be guided down the fibre. An experiment demonstrating this effect has recently been performed by Eric Cornell, Carl Weimann and co-workers at the Joint Institute for Laboratory Astro-physics (JILA) at the University of Colorado.

The idea of an atom pipe opens up some interesting possibilities. The atoms in the fibre can be further manip-ulated by shining light in through the sides of the fibre. Savage has even suggested that this can be used to make a laser of atoms. The output of such a laser would not be a beam of coherent light but a beam of coherent matter, a quantum matter beam. No one really knows just what such a beam would look like, although, as I discuss below, we now know what a pool of coherent matter would be like. The communication revolution that is the Internet is due entirely to the speed with which optical fibres can carry information. The information is passed by small pulses of light which travel, of course, at the speed of light. Coherent matter would travel much more slowly, but could also carry information. While it will never compete with optical com-munication, communication with coherent matter may enable us to realise a truly quantum communication channel, perhaps even a quantum computer, described more fully in the last two chapters of this book.

The ability to trap cold atoms with optical dipole forces suggests another exciting possibility—the optical crystal. The availability of ultracold atoms makes it possible to use standing waves to trap atoms into a regular array. Atoms are cooled and trapped in an MOT and then a standing wave is turned on. For appropriate laser frequencies, the dipole force tends to force atoms to the nodes, or intensity minima of the standing wave. The atom is now spatially confined at the node and thus can only have particular energies, including a lowest non-zero energy. If the cooling is good enough, a single atom can be trapped in this state of lowest energy. We end up with a regular array of atoms localised near the nodes of the standing wave. In three dimensions this can be used to produce a kind of crystal in which atoms are held in place by light alone.

The first such optical crystal in two dimensions was demonstrated in 1993 by A. Hammerich and T. Hansch at the Max Planck Institute near Munich. In the same year a group at the University of Paris led by G. Grynberg demonstrated a three-dimensional optical crystal. Unfortunately these crystals have a lot of gaps (called defects), as most of the nodal points do not contain an atom. No doubt this problem will eventually be overcome, leading to what is, in many ways, an entirely new state of matter—the optical crystal.

Devices and desires

Forcing atoms to behave according to the rules of the quantum theory is an exciting piece of pure science but are there any useful applications? It is early days, but a number of exciting applications have already been suggested and demonstrated in the laboratory, ranging from new semiconductor fabrication to ultra-accurate measurement of gravitational acceleration. On the distant horizon is the possibility of making an atom laser which, instead of producing a highly directed beam of light, produces a directed beam of 'coherent matter'.

The rapid progress in computer technology is due entirely to the ability to manipulate matter at very small scales. Most devices are currently produced by etching specially prepared semiconductor crystals (see chapter 4). This technology represents the culmination of an engineering process, centuries old, by which artifacts are made by 'sculpting' a bulk material. We are now at the threshold of a new technology for a kind of microfabrication in which devices are constructed atom by atom. Can we use light to fabricate new kinds of electronic circuits by directing atoms onto surfaces in a precisely controlled way? A number of groups around the world have been exploring this possibility and the results suggest that indeed atom optics may play a big role in engineering at the atomic scale, a technology best referred to as nanotechnology.

Atom optical lithography

The rapid advances in semiconductor technology in recent decades have been driven by a succession of breakthroughs in lithography, the central fabrication technology of the microelectronics industry. Lithography is an ancient technique based on selective etching of prepared surfaces. In the first stage the material to be etched is coated in some material, the resist, which will protect the underlying surface. This material is then selectively removed in a precise and predetermined pattern. Finally the material is placed in an environment which enables the unprotected parts of the surface to be modified either by removing substrate material or by adding a new layer of a different material. The patterns are usually imprinted by a photographic-like process in which the pattern is projected onto the resist. If the resist can change its chemical properties in response to the light, it can then be selectively removed by further chemical treatment. In fact it is not absolutely necessary to use light. Carefully controlled beams of electrons can also be used to write patterns onto the material, a process known as electron beam lithography.

The size of the pattern is limited by the quality of the optical image formed in the exposure stage. Ultimately this limit is set by the wavelength of the radiation used for exposure. In recent years X-rays, which have a much smaller wavelength than visible light, have been used. Electron beams also enable fine patterns to be written but suffer from a problem, as electrons are charged and thus repel each other if they get too close together. This limits the resolution of the pattern. Furthermore, highly focused X-rays and electron beams can damage the underlying material in unpredictable ways. Low-energy beams of neutral atoms would not suffer from such problems, and appear to be an attractive alternative to conventional lithographic schemes if a method can be found for directing and focusing them onto the substrate. The newly discovered techniques for manipulating neutral atoms with light may be the answer.

Figure 3.6 Atomic lithography

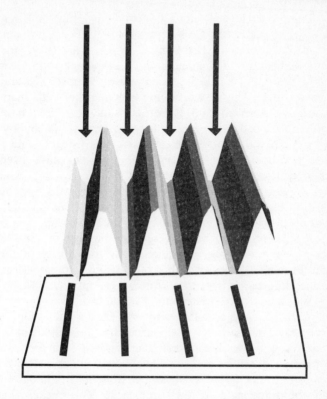

Note: Atoms can be focused into lines on a substrate by standing waves of light.

The first demonstration of the potential for atom optics as a lithographic tool came in 1992 when Greg Timp and co-workers at AT&T Bell Laboratories at Murray Hill, New Jersey, demonstrated that a laser standing wave could be used to focus a beam of sodium atoms into a line on a substrate of silicon. The basic idea is illustrated in Figure 3.6. Atoms are pushed away from regions of high intensity, resulting in a focusing of a broad beam into an array of lines on the surface.

The possibility of atom optical lithography is being actively pursued by the Consortium for Light Force Dynamics, a collaboration between Harvard University, the National Institute of Standards and Technology (NIST) at Gaithersburg, AT&T Bell Laboratories at Murray Hill and Colorado State University. Greg Timp, together with Mara Prentiss at Harvard and Robert Celotta at NIST, have further developed the prospects for atom optical lithography. Recently linewidths of about 46 nanometres (roughly equivalent to about 40 atoms laid side by side) have been achieved. Using standing waves with very deep valleys, and laser cooling of the atoms, even smaller linewidths may be achieved together with high contrast. The potential for atom optical lithography is enormous. Not only can very narrow features be produced but a very large number of features can be written in one pass due to the periodicity of the light forming the lenses. This high degree of parallelism means large patterns can be written very quickly.

We are not restricted to patterns of lines. Robert Scholten, formally at NIST and now in the Atom Optics Group at University of Melbourne, has recently demonstrated how a two-dimensional standing wave can be used to produce a regular array of 'dots' of chromium atoms on a substrate. Each spot is separated from its neighbour by half the wavelength of the light used. This means that something of the order of 1 billion spots per square centimetre can be deposited. If the substrate was moved during the deposition process, more complex patterns could be laid down. That would give 1 billion identical patterns repeated regularly over the substrate. If the resulting patterns were the basis of a transistor or memory switch, this would be a very useful level of integration.

Another way atom manipulation can be used for nanoscale lithography is to modify the resist. Instead of using light to direct atoms to particular regions of the substrate, we can direct the atoms onto the resist. If the atoms can modify the chemical properties of the resist in a manner similar to light-sensitive resists, very fine patterns can be written onto the resist, and transferred to materials such as gold by chemical etching. This technique is being pursued by George Whiteside and co-workers at Harvard University (also part of the Light Force Dynamics Consortium). They are using a self-assembled monolayer (SAM) for the resist. Atoms striking the SAM will alter its chemical properties, making it sensitive to the further chemical processes required for etching. By using the atoms to modify the resist rather than depositing them directly onto a substrate we are not restricted to using only those atoms which will stick to the surface. In addition, by using the right kinds of atoms and resist it may be possible to achieve a strong effect on the resist with just a few atoms, and this opens up another interesting possibility.

The SAM will change if an atom hits it. While light forces give us considerable control over where the atoms land on the resist, to a large extent the impacts are quite random. After all we start with a hot gaseous ball of atoms. Laser cooling and trapping is designed precisely to slow down and limit the motion of the atoms in this hot gas, but a small erratic component remains. Ultimately, for sufficiently cold atoms, we reach a situation where the unavoidable quantum uncertainties in the atomic positions lead to some randomness in where the atoms land on the resist. It might seem that this will place an unavoidable limit on possibilities of atom optical lithography, but in fact atoms which have entered the quantum domain can be manipulated more easily than if they were little billiard balls. The idea is to ensure that the probability for an atom to arrive at some place on the resist is determined, not by the ordinary rules of chance, but by a quantum probability *amplitude*. If we can get a handle on the quantum probability amplitudes directly, we can develop new methods for steering the atoms to particular locations on the screen. The objective is to 'rig' the probability for an atom to arrive at some point by modifying and controlling the

Figure 3.7 Atomic interference

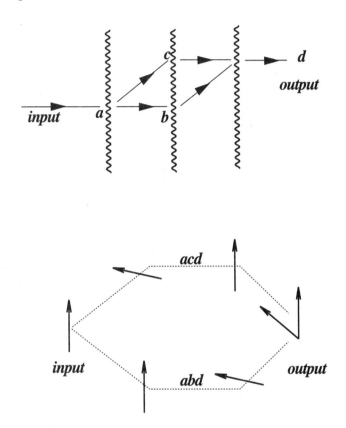

Note: We provide two indistinguishable paths for an atom to get from a to d (top). The probability amplitude for an atom on different paths is not the same (bottom). If the paths are now recombined the probability amplitude of detecting an atom in the original state is the sum of the amplitudes for the two paths to the output. If the paths are not exactly the same, the probability amplitude for an atom on path acd is shifted slightly with respect to that for an atom on path abd, so that there can be partial cancellation of probability amplitudes at the output. This leads to a reduced chance of finding the atom in an unchanged state at the output. Of course, the chance of finding the atom in the other output state (dashed line) is enhanced. In the experiment of Kasevich and Chu, the paths are split by counterpropagating laser fields.

---

underlying probability amplitudes. Such methods are now being pioneered in a number of laboratories around the world.

We can get an idea of what is involved by considering a very simple situation. Suppose we want to place a dotted line of atoms on the resist (see Figure 3.7). One way to do this is to direct a beam of atoms onto the SAM, but insert a 'chopper' to periodically interrupt the beam on its way to the surface. This method is rather wasteful of cold atoms. A better though much more difficult way to do it is to use quantum atoms. The trick is to provide at least two distinct, but unknown, paths for the atoms to reach the SAM. Each path has an associated probability amplitude, represented by a little arrow. The total probability amplitude for an atom arriving at some point is the sum of the arrows for each path considered separately.

If the paths are identical in every respect, so that atoms on each path experience exactly the same forces and travel the same distance, the arrows, representing the amplitude to arrive at some point on the SAM, point in the same direction, and thus reinforce each other to give a high probability to hit the SAM. On the other hand if we can alter one path independently of the other, the arrows will no longer point in the same direction. If we are careful it may even be possible to get the arrows to cancel exactly, so that even though atoms are directed to that part of the SAM, there is no chance at all that an atom will arrive there. Just because the probability amplitudes cancel, the atoms themselves are not 'cancelling', or vanishing from the experiment. They just don't arrive at that part of the screen; they go somewhere else. In this way we use all the atoms we have at our disposal but 'weight the dice' that determines their chances of getting to various places on the SAM.

There are many ways we can modify the paths that the atoms can follow. The simplest way is to put a 'hill' along one path to slow the atoms. The hill, of course, is just some force that we apply to slow the atoms down a bit, and we can use an optical dipole force to do this. Thus we can use laser beams to modify individual atomic paths quite easily. Because of the control we can exert over beams of light, changing their direction and intensity for example, we have a great deal of flexibility in how we modify the probability amplitudes that determine where atoms arrive on the screen.

Experimentalists at Stanford University have recently shown how we can modify the quantum probability amplitudes of moving atoms, and build a very sensitive instrument to measure the acceleration due to gravity. As Galileo discovered centuries ago, two objects dropped from the same place will fall at the same rate. But two objects dropped from different places do not necessarily fall at quite the same rate. It all depends on the local distribution of matter beneath them. The deviations are small but big enough to measure, and can give valuable information as to just what is under your feet. This has important applications to geophysical exploration.

The core of the Stanford experiment carried out by Steve Chu and Mark Kasevich is the atomic fountain. In this device atoms are cooled and trapped in an MOT. By changing the frequency of one of the optical trapping beams, a net force causes some of the atoms to be launched up out of the trap. The launch velocity is quite small, as little as two metres per second, and can be carefully controlled. Typically about ten million atoms are launched vertically. Towards the top of the trajectory the atoms are hit with a laser field. This laser gives a small kick to an atom, making its motion sufficiently uncertain that there are now two possible paths the atoms can follow as they fall. A second laser beam is used to recombine the paths.

The quantum probability amplitudes of each path are not precisely the same and are very sensitive to the acceleration due to gravity. Kasevich and Chu have shown that the acceleration due to gravity can be measured with a resolution of at least three parts in 1000 million and another 100-fold improvement is possible.

To really exploit quantum probability amplitudes with atoms we need to go beyond the Stanford experiment. What we need is a beam of atoms in the quantum domain, in other words we need a beam of quantum matter. This is not going to be easy, but recently some exciting experiments have indicated just how it can be done.

Quantum matter

I am writing this book on a computer and the computer is of course connected to the Internet. Not long ago, as I was typing, I received notification from the mail program that a new message had arrived. I quickly opened up the mailbox to see who it was from. It turned out to be a message from a colleague informing me of a major break-through in the field of atom optics. Physicists in the Joint Institute for Laboratory Astrophysics (JILA) at the University of Colorado had created a new state of matter by cooling atoms to just 20-billionths of a degree above absolute zero. At 10.54 a.m. on 5 June 1995, a bunch of atoms held in a modified optical trap formed a Bose-Einstein condensate—in effect quantum matter. About 2000 atoms had surrendered their individuality and classical distinguishability and crossed the border between the classical and the quantum world.

The JILA team, led by Eric Cornell and Carl Weiman, were the first to succeed in a quest being pursued by many laboratories around the world. They used atoms of the element rubidium, trapped in a modified MOT called a TOP (for time-averaged orbiting potential) trap. Most of the properties of this new state of matter, sometimes called coherent matter, are still unknown. We cannot picture this state as simply a bunch of little billiard balls. It is rather difficult to picture at all. However, we do know, in principle at least, how to describe it using the quantum theory. A great deal of theoretical and experimental work is now under way to understand just what has been produced. But one thing is clear, it opens up some exciting possible applications in atom optics.

We are still a long way from having a beam of quantum matter. In the JILA experiment the Bose-Einstein condensate formed in a small pool at the bottom of the trap. What we need is some way of producing a directed beam of coherent matter. In effect this would be a laser for matter, as a beam of Bose-Einstein condensate would be the analogue of the way light condenses into a laser beam. A number of ingenious methods have been suggested to produce such a matter laser, sometimes called a 'boser', but much work remains to be done. I am confident that sometime in the next few years one of these schemes, or perhaps some as yet undiscovered scheme, will be made to work.

Atom optics represents an alternative technology for manipulating single atoms to that described in chapter 2. In some ways it is even more interesting, as the atoms are forced to behave not as small billiard balls but as true quantum objects. The ability to control the motion of massive quantum objects is likely to lead to as yet unimagined applications. Very recently (May 1995) a group at the University of Hanover demonstrated optical imaging with quantised atomic beams. In this experiment atom optics is seen to be following the path that electron optics travelled many decades ago. This field is barely ten years old. We can expect to see some exciting developments in the near future and it probably won't be long before the first Nobel Prize is awarded in this subject.

CHAPTER FOUR

QUANTUM NANOCIRCUITS

The computer revolution is the direct outcome of the successful quest to fabricate electronic circuits on smaller and smaller scales. The journey down to microcircuits, however, will eventually sail beyond the classical horizon. To continue on the present course will deliver circuits so small and fast that they will self-destruct through an inability to dissipate heat faster than it is generated. These physical limits are discussed further in chapter 6. Beyond this horizon lies the land of the quantum. To explore this new realm we must abandon forever familiar classical paths, and walk the multibranching, probabilistic paths of quantum mechanics. In this chapter I will describe new kinds of fabricated semiconductor circuits which are already exploiting quantum rules to deliver new devices. In these systems fundamental charge carriers, such as electrons, are forced to exhibit their quantum nature. Classical laws, such as that discovered by Georg Simon Ohm in 1826, are broken and new quantum rules must be discovered to describe the resulting circuits.

Electricity by gaslight

In the gaslit laboratories of the nineteenth century, physicists and chemists laboured to understand the nature of electricity. The seventeenth and eighteenth centuries laid the foundations for the edifice of classical physics, building on the works of Newton. In the nineteenth century, the age of electricity, the age of Faraday and Maxwell, the second wing of classical physics was completed—the theory of electromagnetism. In one such laboratory, that of Michael Faraday, in 1833 a surprising result was observed. When heated, the poor conductor silver sulphide became a better conductor. This was an unexpected result as all metallic conductors showed precisely the opposite behaviour, becoming worse conductors when heated. With the benefit of hindsight we now know that what Faraday observed was the defining characteristic of a semiconductor, a material central to the technology of the twentieth century.

Maxwell's theory of electromagnetism was entirely mathematical. It gave relations between observed quantities, such as current and charge, and inferred physical quantities such as the electric and magnetic field. But the theory gave no account of electromagnetism in terms of the constituents of matter. It was not until the work of H. A. Lorentz in 1895, and the introduction of the concept of the electron, that a microscopic account of electrical and magnetic effects could be given. By the early 1920s it was generally accepted that electrical conduction was the result of electrons moving almost freely between the ionic cores of metal crystals. In 1928 Felix Bloch, a student of Heisenberg's in Leipzig, applied Schrödinger's quantum mechanics to electrons in solids. As a result of his work the modern theory of electrical conduction was born, and with it an explanation for the curious observations made by Faraday in 1833.

Electron pinball

In a single atom of a metal such as sodium, the electrons are held in place by the strong electrical attraction of the positive nuclear core. When sodium atoms get together to form a metallic crystal, the outermost electrons are released from the coulombic prison that binds them to a particular atomic site. These electrons are free to roam throughout the crystal, leaving behind a positive ionic core. The result is a noisy crowd of electrons milling around the almost rigid, positively charged ionic cores of the crystal. Of course, the number of positive charges and negative charges are exactly equal, so the crystal as a whole is electrically neutral.

The electrons are not quite free. Firstly, as the electrons are all negatively charged, they must scatter off each other incessantly. Secondly, they must still feel the effect of the positively charged ionic cores. Imagine a flat dimpled surface, a regular array of small depressions, with ball bearings rattling across it. The dimples are not deep enough to capture any particular ball bearing, but they will slow it down slightly and change its motion. So does an electron feel the electric attraction of the ionic cores of the crystal.

In a real metal the ionic cores are not rigidly held in place but can wobble around an equilibrium position. This makes the collisions of electrons and ionic cores not quite elastic and an electron can lose or gain energy from the wobbling ionic core at each collision. Eventually an equilibrium is reached; on average electrons neither gain nor lose energy. This is the classical picture of electrons moving in a metallic wire. If we connect the ends of the wire to a battery the electrons begin to accelerate towards the positive terminal of the battery. They do not get very far before they collide with an ionic core and lose whatever energy they gained. As a result of the collisions, the ionic cores wobble a little more violently, which appears to an outside observer as a heating of the wire. This process of acceleration followed by collision leads to an eventual steady average flow of electrical current through the wire.

If we externally heat the wire, we cause the ionic cores to move more rapidly, which increases the chance that an electron will suffer energy loss when it collides with an ionic core, and thus decreases the overall conductivity of the wire. Imagine trying to play a game of pinball in which the bumpers randomly jump from place to place. If the jumps are large and frequent enough the ball will take a very long time to get to a flipper at the bottom.

In contrast to a metal, the electrons in a non-conducting

crystal are held tightly to the atoms forming the crystal. If we now connect the ends of the crystals to a battery we may push the electrons slightly to one side of each atom (known as polarisation) but the electrons remain tightly bound. If there were only metals and insulators this theory would look pretty good. But when it comes to explaining the observation by Michael Faraday, it does not do so well. To explain this we need the quantum story of electrical conduction. Fortunately we already have enough quantum rules to figure out roughly what might happen.

An electron is an elusive creature. When it is isolated from an atom it appears for all the world like a little charged particle. But when it is held in the grip of electrical forces inside an atom it behaves altogether differently. It is then subject to the rules of quantum theory and its location is probabilistic and determined by probability amplitudes. The result of restricting its motion, according to the quantum theory, is also to restrict the energies it may have. Thus an electron in an atom is restricted to particular energy levels.

In a metal, an electron lies somewhere between the two extremes of an isolated free particle and a particle held in place by strong forces. The result is behaviour that lies between that of a free particle which can have any energy at all, and a trapped particle which can only have particular discrete energies.

To begin our quantum theory of a metal we note that whatever happens inside, the electron is at least confined within the surface of the metal. The origin of this confinement is also electrical. If an electron tried to leave a metal it would see behind it a positively charged block and soon be pulled back into place. These strong surface forces were discussed in chapter 2, and can be overcome through the subtle quantum effect of tunnelling. So at the very least an electron is only free within the confines of the metal. This gives it a pretty big range, but nonetheless it is a restricted range and according to the rules of quantum mechanics, it must have a restricted energy. However, because the extent of the confinement is so large, the gaps between allowed energy levels are very small, so small in fact that for all practical purposes we might as well say that all possible energies are allowed.

Recall that the probability of finding a free particle at a particular position is determined by a probability amplitude. In chapter 1 this amplitude was represented as a little arrow, which rotates as we move from one position to another. How much and in which direction it rotates depends on the momentum of the particle in question. At any point a particle can be traveling in any of six mutually perpendicular directions (up–down, left–right, front–back). If all we want is the total amplitude to find a particle at a particular position, independent of its direction of motion, we must first add the amplitude for each of the six ways a particle can be moving at this position. Now it is possible that the arrows corresponding to these six possibilities can cancel exactly to give no chance of finding a particle. This is precisely what happens at the surface of the metal. As we move from one face of the metal to another the arrows must all rotate by just the right amount to cancel exactly at the other side. This requirement restricts the angles through which we can rotate the arrow and, as the angle is determined by the momentum, the momentum is restricted as well. The energy is proportional to the square of the momentum, so the energy is also restricted.

But even within the confines of the metal an electron is not entirely free. It still feels electrical forces at the ionic cores of the crystal. As it passes by an ionic core it can suffer a change in momentum as the electrical forces exerted by the core take effect. This makes it difficult to think of a particle moving with a fixed momentum, but we can still consider particles of fixed energy.

The energy of the particle comes in two forms, an energy of motion proportional to the momentum squared and a potential energy proportional to the electrical force exerted on the particle. As an electron slides by an atomic core, some energy of motion gets changed to potential energy. The particle slows down, but the overall energy remains the same. However, because the momentum changes with every encounter with an atomic core, the degree by which we rotate the probability amplitude as we move around inside the crystal becomes rather indeterminate. Fortunately because the ionic cores are arranged in a symmetric crystal lattice, we can work out the overall effect on the probability amplitude for position measurements. This is precisely what Felix Bloch did in 1928, only three years after Schrödinger published his famous equation which enables us to calculate probability amplitudes.

How we adjust the free particle probability amplitudes depends on how much overall momentum the electron has. For most values of momentum the adjustments we need to make as we move from one side of the crystal to the other are minor, and it is easy to arrange a complete cancellation on every face of the crystal. For other values of the momentum, the changes in the angles of rotation induced by the ionic cores make it impossible to get complete cancellation at the edge of the crystal. The energies corresponding to these cases simply cannot occur. This happens over a wide range of momentum values and is called a gap. Between the gaps we get broad bands of allowed energies. Within a band there are a large number of closely spaced allowed energy levels.

The electrons roaming amidst the ionic cores of a metal are almost free, but their energies are restricted to bands of allowed values separated by gaps. Within an allowed band the energy is determined by the momentum, but is not simply proportional to the square of the momentum as for a free particle. However, in some substances this is a good approximation at least over some range of momentum.

Electrons obey another rather unusual quantum law with far-reaching consequences. The rule, called the exclusion principle, says that no two electrons can occupy the same allowed energy state. This means that the electrons in the allowed energy bands all have slightly different energies. Electrons gradually fill up the closely spaced energy levels within a band, similar to the way electrons fill energy levels in an atom. Every added electron must go to a state of higher energy. How full a band is depends on which material we are considering. In a metal the band with the highest energies is not completely full. An electron can then get a little bit of extra energy, perhaps provided by a battery, and move a little higher in the band, resulting in a current flow. Such a partially filled band is called a conduction band for obvious reasons. Once an electron begins to move it still can bounce erratically off the ionic cores, losing some energy to heat in the process. So the classical description of resistance and conductivity is unchanged.

Suppose a band is completely full, with the next available energy in a higher band, separated by a band gap. If we try to supply extra energy, but less than what is required to bridge the gap, the electron simply cannot take it on. Nothing changes and no current can flow. Such a substance is an insulator. A filled band such as this is called the valence band.

There are some materials for which the gap between the valence band and the conduction band is not too large. In fact it may be small enough that we can supply the bridging energy simply by heating the crystal. Now some electrons will get enough energy to move across the band gap to the conduction band where they can move, thus resulting in an electrical current. The hotter it gets the more electrons make the jump and the greater is the electrical conductivity. So 95 years after Faraday's experiments on silver sulphide, the correct explanation was implicit in the thesis work of a young German PhD student, Felix Bloch.

Bloch subsequently went on to even greater achievements in physics, including research in nuclear energy at Stanford and Los Alamos and later working to devise countermeasures against radar at Harvard. After the war he returned to Stanford and pioneered the field of Nuclear Magnetic Resonance (NMR), which today provides the basis for one of the most successful non-invasive medical diagnostic tools, MRI or magnetic resonance imaging. He was awarded the Nobel Prize in 1952 for his work in NMR.

The silicon century

There are some who would claim that the most decisive scientific event of the twentieth century took place in the English city of Manchester in 1919, in the laboratory of Sir Ernest Rutherford. Newspaper headlines summarised the result—Rutherford had split the atom. Considering what happened in Hiroshima 36 years later, there is clearly some truth to this claim. But in 1945 the century was not even half done. Subsequent events have indicated that nuclear energy for both peaceful and military ends is nowhere near as beneficial as had been hoped. The claim of most important scientific breakthrough of this century may ultimately be substantiated for a much quieter event which took place in the Bell Laboratories of AT&T in New Jersey in 1948. This event did not reach the headlines. The *New York Times* devoted only 4.5 column inches (11.5 centimetres) on page 46 and the story was preceded by the apparently more momentous news that the radio station WNEW would soon broadcast traffic information (Braun and Macdonald 1982). The article said:

> A device called a transistor, which has several applications in radio where a vacuum tube ordinarily is employed was demonstrated for the first time yesterday at Bell Telephone Laboratories, 463 West Street, where it was invented.

It seems to me that the consequences of the Bell experiment have affected the lives of ordinary people far more profoundly than the consequences of splitting the atom. Let us hope this continues to be the case in the coming century.

The discovery of the transistor was something of an accident, but fulfilled a dream of electronics engineers to make a small solid-state amplifier to replace the bulky, energy-hungry valve amplifiers then in operation. The three Bell scientists subsequently awarded the Nobel Prize (in

1956) for this discovery were John Bardeen, Walter Brattain and William Shockley. Shockley's original suggestion for such a device did not work. It was while Bardeen and Brattain were attempting to understand why it didn't work that the point contact transistor was discovered. The heart of this device was a small crystal of germanium, a semiconductor.

By the late sixties many companies were producing single transistors packaged in a variety of ways. In February 1959, Jack Kilby of Texas Instruments filed a patent describing how two transistor circuits could be fabricated on a single piece of germanium. But the crucial innovation took place at Fairchild Semiconductor, where Robert Noyce invented the planar process of integration. This technique is more suited to silicon than germanium and led directly to the dominance of this element in the semiconductor industry. Noyce's approach would eventually realise an entire computer central processing circuit integrated onto a single chip of silicon.

At about the same time that Fairchild introduced the planar process for circuit integration, J. R. Schrieffer suggested that confining electrons to junctions between different materials would lead to phenomena in which the quantum nature of the electron would be apparent. The idea could not be fully realised until the advent of the new fabrication technique of epitaxial film growth became available. This process is based on the growth of crystal structures by laying down single atomic layers. By carefully controlling the kinds of atoms laid down a whole class of artificial crystal structures can be developed in which the energy bands are tailored rather than simply served up ready-made by nature. As such crystals are made up of more than one substance they are called heterostructures.

2DEG, 1DEG, 0DEG

Heterostructures are the raw material for quantum nanocircuits. The crucial innovation is the ability to confine electrons to a thin plane between two different materials, typically gallium arsenide (GaAs) and an alloy of gallium arsenide and aluminium (AlGaAs) (Bate 1986). The technique of epitaxial growth enables atomically flat layers of these material to be deposited one on top of the other. At the interface electrons are trapped in a layer so thin that they are subject to a quantum restriction of energy levels, the so-called 'quantum-size effect'. Parallel to the interface the electrons behave just as freely as they would in a bulk semiconductor. The result is a flat electron gas known as a 2DEG (for two-dimensional electron gas). In the plane the electrons are as free as atoms of air, but parallel to the interface the electrons are held in a vice of electrical force. In this direction quantum speed limits apply.

But we needn't stop there. By a variety of techniques, one of which I will describe below, we can confine the electrons again. The same techniques which have led to the ability to write integrated circuits at a scale of one-millionth of a metre can be used to write confining patterns onto the 2DEG. If we constrain one other dimension, we end up with electrons confined to move on a line. This is a 1DEG, better known as a quantum wire. Now the quantum rules restrict the energies for motion in two directions. If we confine the gas in the only remaining free direction, we end up with a quantum dot or 0DEG (although the gas analogy is beginning to look a bit strained by this stage). The result is a little box which may contain as few as 100 electrons. Inside the box the electrons rattle around as if completely free. (In fact they are not quite free, the box is much larger than the underlying atomic lattice so the electrons still see the periodic forces due to the ionic cores. These forces effectively make the electrons appear to be a bit less massive than they really are, but for all practical purposes they are still essentially free.) The box can be so small that the quantum rules restrict the energies for motion in every direction resulting in fixed discrete energy levels. The result is a kind of artificial atom, but one that can be built to order, a DIY atom.

William Shockley's original idea for a semiconductor amplifier, a field effect device, was not the basis of the first successful transistor. Indeed it might be said that the transistor was the direct result of failing to realise a field effect amplifier (Braun and Macdonald 1982). As it has turned out, the idea is sound once a number of problems are overcome and indeed this kind of transistor forms the basis of almost all integrated circuit chips today. The basic idea is very simple. Take a slab of semiconductor and make it one plate of a capacitor, while the other plate is a metal conductor. A voltage applied to the metal plate, called the gate, can either induce additional charge carriers into the semiconductor (as in a MOSFET or Metal Oxide Semiconductor Field Effect Transistor) or deplete charge carriers already present (as in a MESFET or Metal insulator Semiconductor Field Effect Transistor). In this way the current in the channel formed by the semiconductor can be controlled by applying voltage to the gate. Silicon MOSFETS of less than one-millionth of a metre can now be fabricated on a single chip, and the sale of such devices dominates the multi-billion-dollar trade in semiconductor chips today.

Our story will turn to the other kind of FET, based on the depletion of charge carriers in the semiconductor. This kind of device is appropriate for the heterostructure devices such as GaAs/AlGaAs in which a 2DEG may be formed. The basic idea is to form metal gates over the top of the plane containing the 2DEG. By applying a voltage to the gate, electrons can be forced out of the 2DEG, thus creating barriers to current flow. This technique was first demonstrated in 1986 by T. J. Thornton, M. Pepper, H. Ahmed, D. Andrews and G. J. Davies in the Cavendish Laboratory at Cambridge University, and goes by the name of surface-gate technology. Crucial to this breakthrough was the development, throughout the early eighties, of techniques to place carefully structured metal patterns onto semiconductor heterostructures. The result has been a new era in semiconductor physics in which length scales have decreased from millionths of a metre to one-thousand-mil-lionth of a metre (a nanometre) or a shift from microscopic to nanoscopic physics.

Surface-gate technology opens the way to a quantum circuit. However, it is not just a simple matter of making devices smaller. The life of an electron in a 2DEG is not necessarily determined by pure quantum laws. While the electron is very nearly free to move in a plane parallel to the heterostructure interface, it still has to contend with the underlying ionic cores of the crystal structure. In addition, this crystal structure may contain impurities, a foreign atom, or even breaks and dislocations in the regular arrangement of the lattice. While we can carefully model the effect of the regular array of ions in the lattice, we have no knowledge at all of the details of defects and impurities. Furthermore, at finite temperature the ionic cores are wobbling around in a random way which we can only describe statistically. Were it not for these complications we could use Schrödinger's equation to assign probability amplitudes for an electron at any point in the plane for any electron speed. But if an electron encounters a defect or impurity or bounces off a randomly moving ionic core, we have no way of knowing how to modify the probability amplitude. Worse still, different electrons encountering different defects will suffer changes to their probability amplitudes which are entirely unrelated to each other. The net effect is to randomise the probability amplitudes at each encounter with a disturbance. As a result of this randomisation, subtle quantum effects based on cancellation of amplitudes get washed out and classical behaviour sets in.

There are two things we can do to prevent a randomisation or, as it is technically known, phase-destroying collision. We can try and make ultra-pure samples in which defects and impurities are carefully controlled. This is exactly what is done, and indeed the artificial crystals grown to form such devices are probably the most pure and perfect artificial constructions ever built. The only way to reduce the effect of random lattice vibrations is to cool the devices.

Typically liquid helium temperatures are used, a few degrees above absolute zero. These methods do not eliminate phase-destroying collisions, they merely reduce the chance of such a collision. However, if the device is small enough and the speed or mobility of the carriers great enough, the probability for a phase-destroying collision anywhere in the device can be reduced to near zero. Thus the essential need for nanoscopic dimensions. Quantum nanodevices are very cold, extremely tiny, near-perfect electrical devices.

Quantised conductance

In 1988 two groups showed that quantum mechanics would dramatically modify current flow in a nanoscopic circuit. Both groups used a technique referred to as a split-gate (Taylor 1994) (see Figure 4.1). A metal strip is deposited over a GaAs/AlGaAs heterostructure bar. The strip extends all the way to the edges of the bar, but in the centre contains a small gap, thus a split-gate. A negative voltage applied to the metal gates depletes the 2DEG beneath to form a channel or wire. The result is two wide sections of 2DEG joined by a narrow constriction, sometimes referred to as a quantum point contact (QPC). Current is injected into the device through contacts on the wide regions of electron gas. The objective now is to vary the width of the channel and measure the resistance or rather the inverse of resistance, a quantity known as conductance.

At first sight it may appear that this requires manufacturing a whole bunch of devices all with slightly different channel widths. The devices would need to be identical or nearly so and this would not be easy. Fortunately there is a much easier way to do it. The actual depletion region is wider than the gates and depends on the voltage. Thus the width of the constriction can be varied by changing the gate voltage.

Before going on to describe the two seminal experiments on QPCs, I want to turn the clock back to the nineteenth century, to 1826 to be precise. Georg Simon

Figure 4.1   A split-gate device

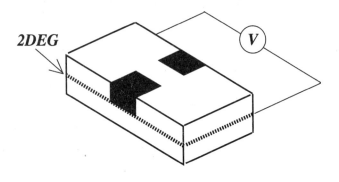

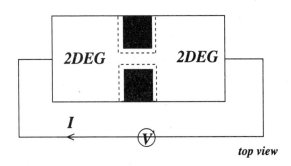

top view

Note: A two-dimensional electron gas (2DEG) is confined to a flat narrow layer between two different semiconductor materials. Metal electrodes (black blocks), called a split-gate, are deposited on the surface. By applying a bias voltage to these electrodes, electrons can be depleted in the region of the gas beneath. The depletion region covers an area larger than that of the electrodes (dashed region in top view). This area can be made larger by increasing the bias voltage. An additional voltage, V, is used to drive electrons through the narrow channel formed by the depletion regions.

Ohm was a teacher in the gymnasium in Cologne. Labouring quietly with homemade apparatus, Ohm began to study the question of electrical conductivity. This may seem a rather modest undertaking, but you need to place yourself in this epoch. The scientific study of electrical current was only a few decades old and conducted in large part by a group of gifted men with both the leisure and the money to pursue their ideas. Ohm had neither. The work of Faraday and Maxwell lay decades in the future. Despite his obvious lack of equipment and time Ohm was able to establish his famous law, now taught to every child at an early age.

In fact Ohm discovered two laws. The second and much the better known says that the current that flows in a wire increases proportional to the voltage difference between the ends. The constant of proportionality is called the conductance. We usually teach this the other way around and say that the voltage drop across the wire divided by the current is always a constant called the resistance. It is the same result provided we identify conductance as the inverse of the resistance.

Now any student can repeat this experiment in a high school laboratory. They simply take off the shelf a battery, a current meter and a few bits of wire. Ohm of course had neither a ready-made current meter or battery. He had to proceed far more ingeniously. To generate a constant source of voltage he used a phenomenon discovered, in 1821, by Thomas Johann Seebeck, a wealthy Russian working in Berlin. Seebeck found that if a circuit consisting partly of copper and partly of bismuth is heated, a current will flow. He called it thermomagnetic current. We now call it thermoelectricity. To measure the current, Ohm used Oersted's discovery, in 1819, that a current will deflect a compass needle. By connecting the needle to a torsion balance, Ohm was able to deduce the relative change in the current by looking at deflections of the needle.

The first law Ohm discovered was this: the conductance depends on the material, but for any given material it increases proportionally as the cross-section of the wire increases and decreases as the length of the wire increases.

Not long after his famous experiments, Ohm published a theoretical derivation of his results. At that time no one considered the flow of electric current to be due to the stop-start diffusion of electrons through a crystal. In fact the usual explanation was based on an analogy between electric current and the flow of a fluid. The voltage across a circuit was analogous to a pressure difference between the ends of a pipe. For a fixed voltage, increasing the cross-sectional area of the pipe obviously increases the fluid current, thus the conductance of the pipe increases as the surface area increases. Even today this is a helpful analogy for beginners. In the experiments on the QPC this analogy is hopelessly inadequate.

In a collaboration between Philips Research Laboratory in the Netherlands and the Philips laboratory in England with Delft University of Technology, B. J. Wees and his colleagues constructed a QPC in 1988. In this device the maximum width of the channel was about 250 nm, but could be steadily reduced to zero by the application of a negative bias voltage to the gates. Current is injected into the device by the application of a potential across the large 2DEG regions either side of the channel. The total current through the circuit is then monitored as the width of the channel is changed, and the resistance of the circuit is calculated as the ratio of the applied voltage to the current that flows. From this we deduce how the conductance will vary as the width of the channel is changed.

The results of the Dutch group are indicated in figure 4.2. Two aspects are striking. Firstly the conductance does not increase in simple proportion as the width of the channel is increased. Clearly apparent are distinct steps, between which changing the width of the channel makes no difference at all to the conductance of the quantum wire. Similar results were obtained by Michael Pepper's group at the Cavendish Laboratory in Cambridge, at about the same time. These steps in the conductance are now referred to

Figure 4.2   Quantised conductance

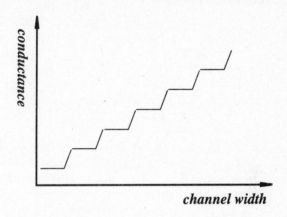

Note: (a) An idealisation of the results of experiments on a quantum point contact (QPC), one of the first indications of quantised conductance in a quantum circuit. As the width of the channel is increased, more and more electrons have an energy matching one of the allowed energies in the channel. This causes the current, and thus the conductance, to increase in steps. The steps are an indication that motion in the channel is restricted to a discrete set of energies. (b) Shown are the results of the first Australian nanostructure measurements on quantised conductance made by Richard Taylor and Richard Newbury at the University of New South Wales. The width of the gate was decreased by decreasing (more negative) the bias voltage. Thus the horizontal axis corresponds to *decreasing* channel width.

as quantised conductance, and the step size is proportional to the square of the charge on the electron and inversely proportional to Planck's constant. The corresponding resistance step is about 10 kΩ. In this phenomenon the quantum speed restrictions are imposed on a real laboratory electrical current.

The results of the two QPC experiments had been predicted many years before by Rolf Landauer, a physicist at IBM Thomas J. Watson Research Laboratories in New Jersey. Landauer took the novel approach of describing electron transport in solids in terms of the fundamental probability for an electron to be transmitted through a channel. This was, at the time, a controversial way to view conductance, and for many years the Landauer formulae for conductance (we shall meet another formula derived by Rolf Landauer in chapter 6, a formula with far-reaching consequences for the information age) was the subject of some debate. As it turns out all that was required was a system in which the strict assumptions of quantum mechanics could be realised. The QPC results in 1988 settled the issue. Landauer's approach has led to something of a renaissance in the study of current transport.

We can easily get at the basic explanation of quantised conductance using the simple quantum rules I presented in chapter 1. The momentum of an electron in the channel is the sum of a component down the channel and a component across the channel. The motion down the channel is not restricted at all, but the voltages on the split-gate restrict the motion of the electron in the cross direction. The quantum rules then tell us that the momentum in the cross direction is restricted to particular values. This is like a quantum speed limit that says you can travel *only* at 40 km/h, 60 km/h and so on, but *never* at any other speed. The particular momenta that are allowed depend on the width of the channel.

Imagine we are looking end on down the channel so that we can only see the crossways motion. The probability amplitudes for electrons moving in opposite directions across the channel must cancel at both walls. As explained in chapter 1, this can only happen if the crossways momentum of an electron is carefully matched to the width of the channel. Now consider the fate of an electron trying to enter the channel. Unless its momentum, in a direction across the channel, exactly matches one of the posted quantum limits, it cannot enter the channel. Unlike a classical billiard ball, it bounces straight back from the mouth of the QPC.

For a given width there are a number of allowed energies, and these allowed energies decrease and bunch up as the width is decreased. This is indicated in Figure 4.3. Each curve shows how each allowed energy level changes as the width changes. I have also drawn a horizontal band on this figure which represents the energy range of electrons in the 2DEG to the left of the channel. For a given width, such as point A in the figure, it may turn out that no electron in the 2DEG has an energy allowed for the channel. In that case no electron enters the channel and no current flows. As the width increases, eventually one level drops within the range of electron energy, point B, and a current can flow. The conductance now takes a sharp step up. Further increasing the width has no effect on the conductance. As the curve dips below the region of allowed energies more electrons can satisfy the quantum speed limits, but all electrons contribute the same conductance. The conductance will then remain unchanged as the width is changed until the next allowed level falls within the range of available energies, point C1. Of course the electrons of lower energy continue to match the requirements for the first energy level, point C2. The conductance takes a step up as a second class of higher-energy electrons enter the wire. The result is a series of steps and plateaus in the conductance as the changing width opens the channel to more and more classes of electrons.

Landauer's formula for quantised conductance in a QPC is a more quantitative version of this basic argument. It does this by computing the probability for an electron to enter the wire using quantum probability amplitudes. For motion confined to a region of space, the ways in which the amplitudes combine ensure that only particular electron energies will be transmitted with significant probability. Quantised conductance in the QPC is dramatic confirmation that the quantum rules for probability are right. In the QPC we see for the first time how quantum circuits will behave.

101

Figure 4.3   Principles of quantised conductance

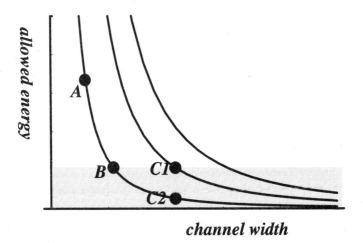

Note: An indication of why steps occur in conductance of a quantum circuit as the width of the quantum channel is varied. The dark lines indicate how the allowed energies in the channel vary as the width of the channel is increased. The shaded region represents the actual energies possessed by the electrons in the 2DEG. Only when the curve representing an allowed energy falls within range of the electron energies can those electrons pass through the quantum point contact. At point A, the curve is too narrow, and no electrons have an energy which matches this value. As the width of the channel is increased, the allowed energy in the channel decreases until point B is reached. At this point the allowed energy matches the energy of some electrons in the 2DEG. These electrons can move down the channel contributing to the current, and the conductance rises in a sharp step. The conductance will remain at this value until the next available energy falls within range of the energies possessed by electrons in the 2DEG, point C2.

Ohm's laws for conductance fail, and we require the new theory of quantised conductance pioneered by Rolf Landauer. This leads to some interesting modifications of basic circuit concepts.

Take the case of resistors in series. Soon after students learn about Ohm's law we teach them how to calculate the total resistance of a circuit containing two or more resistors connected in series or in parallel. For two resistors connected in series, the total resistance is the sum of each resistance. Suppose instead of two ordinary resistors connected head to tail, we connect two QPCs of different widths in series, so that there is a smooth transition from one QPC to the next. What is the total resistance of this quantum circuit? The answer is, the total resistance of the pair is given by the largest resistance of the two. Our students will need a whole new set of rules when they encounter quantum circuits in the nanotechnology labs of the next century.

Quantum dots

Surface-gate technology shows considerable promise in taking us towards quantum circuits. A fairly likely component of a future quantum circuit is the quantum dot. A quantum dot is the natural end point of the quest to fabricate devices at smaller and smaller scales. The interest here stems in part from the potential for quantum dots to operate at the level of a single electron . . . which is surely the ultimate limit for an electronic device. Quantum dots have also been suggested as a component to realise a quantum computer, discussed in more detail in chapter 6. A quantum dot is really an artificial atom (Kastner 1993) but hundreds of times bigger. Instead of a positively charged nucleus holding the electrons in place, heterostructures plus surface-gate biases trap electrons in very shallow pools, which may contain as few as a dozen (or less) electrons. There are other ways to make quantum dots, but I will confine myself to surface-gate methods in 2DEGs. Surface-gates can now be patterned onto a chip to form

Figure 4.4   A quantum dot using QPCs

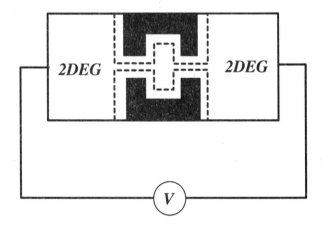

Note: An indication of how split-gates may be patterned to form a box-like confining region for electrons in the 2DEG. The result is two QPCs in series, with a 'quantum dot' between them. Because the depletion region can extend beyond the confines of the metal electrodes, the quantum dot can be isolated from the 2DEG outside the confining region, by pinching-off the QPCs. The only way electrons can enter the dot is by quantum mechanical tunnelling.

---

quantum dots joined to the wider sea of the 2DEG by QPCs. Such a pattern in shown in Figure 4.4. The degree of control afforded by surface-gate technology enables an artificial atom to be made to order. Unlike a real atom, where the constraining force felt by the atom is predetermined by the structure of the nucleus and almost unalterable, the constraining forces in an artificial atom can be carefully controlled. Even the geometry of the constraining forces can be controlled by using various surface-gate shapes.

Quantum mechanics was born when it was realised that electrons bound to real atoms could not reside in orbits with arbitrary energy. For real atoms, the confinement in space is very tight and consequently the gaps between allowed orbits are quite large. Photons of light can be used to provide the energy necessary to shift the electrons from low- to high-energy orbits. As postulated by Einstein in 1905, the energy of a photon is proportional to its frequency. It then appears that atoms will only absorb light at particular discrete frequencies. The set of frequencies absorbed, or absorption spectra, is a signature of a particular atom, and long before the quantum explanation for this phenomenon was discovered chemists had used spectroscopy to identify elements.

Quantum dots are much bigger than real atoms and consequently the gaps between the allowed energies of the electrons are much smaller. It is not currently feasible to make the dots small enough that the gaps between the levels can be probed with light. However, for quantum dots in nanodevices we have another way to proceed. We can use the current to add or remove single electrons from the dot, one at a time. When we try to do this we encounter the essential fact that makes electric current *not* equivalent to the flow of a fluid . . . the charge comes in a smallest unit, equal to the charge on the humble electron. The charge on the electron is the minimum amount of charge we can add or subtract from a quantum dot. The interplay between quantisation of charge (often referred to as the Coulomb blockade effect) and quantisation of energy in the dot leads to very interesting circuits that suggest novel applications. When a magnetic field is added to the picture, the possibilities are seemingly endless. I can provide but a small sampler of this technology, which is only a few years old and holds great promise for both fundamental understanding of quantum theory and new devices.

The SET

If a quantum dot represents the end point of microfabrication trends, the single electron transistor, or SET, represents the ultimate current control device. The objective here is to control an electrical current a single electron at a time. The possible application to digital logic circuits is obvious, with the presence or absence of an electron corresponding to a 1 or 0 respectively. A variety of mechanisms have been proposed to achieve this, but the underlying principles remain the same. To understand how an SET works we must consider the Coulomb blockade effect.

Not long after we introduce students to resistors and Ohm's law, they encounter the next simplest circuit element, the capacitor. A capacitor is simply a means to store charge. The simplest version is a parallel plate capacitor in which two conducting plates are placed close to each other and separated by an insulator (such as air). If each plate is connected to the opposite terminals of a battery, electrons flow onto one plate and off the other plate until the electric repulsion of accumulated electrons on one plate prevents other electrons from joining them. How many electrons we can get onto the plate depends on the geometry of the plates and the voltage supplied by the battery. The ratio of charge to voltage is a constant called the capacitance. The capacitance depends on the geometry of the device used to store charge and, needless to say, the smaller the device the smaller the capacitance.

Once the plates are charged no current will flow on average, but while charging a current flows. We must supply energy to separate charge, and once the current stops flowing we say that this energy is stored in the capacitor. How much energy is stored is inversely proportional to the capacitance. A small capacitance requires a lot more energy to separate a given amount of charge than a large capacitance. This can ultimately be attributed to the fact that charges on a small capacitor are rather crowded, and thus the repulsive forces between them are greater.

It is now possible to make nanoscopic capacitors, as I will shortly explain. These devices are very small and thus have very small capacitances. It may only be possible to store a tiny amount of charge on such a device. Of course, the minimum amount of charge we can transfer at any one time is a single electron. Charge is not continuous but, positive or negative, it comes in little discrete packets. How much energy does it take to get a single electronic charge onto the capacitor? Not much for an ordinary garden variety capacitor, but for a nanoscopic capacitor the energy cost of getting a single electron onto the device may be substantial, at least compared to the energies of the rest of the electrons in the circuit. If the energy cost to load a single electron onto a nanoscopic capacitor is large compared to the energies of the electrons in the leads, the capacitor cannot be charged. This is called a Coulomb blockade.

The Coulomb blockade is not a quantum mechanical effect. It is due solely to the fact the electric charge comes in a minimum size. However, when it is combined with quantum mechanical tunnelling, discussed in chapter 2, we have the basis for a number of novel quantum devices including the single electron transistor and the electron turnstile.

Such a device was fabricated by a group at MIT led by Mark Kastner in 1990, using split-gate technology. The essential component is a quantum dot formed by isolating a region of 2DEG from the surrounding two-dimensional sea by pinched-off quantum point contacts. Classically no electron has sufficient energy to penetrate the quantum dot, as they do not have enough energy to overcome the repulsive forces induced by the surface-gates. However, electrons can quantum mechanically tunnel into and out of the dot, provided they have an energy which matches one of the allowed energies in the dot.

The dot, of course, forms part of an external circuit. The barriers, formed by the pinched-off QPCs, look to the circuit just like insulators and thus the dot appears like an isolated island on which we can store charge. If the dot is very small, the Coulomb blockade energy may be greater than the energy of any electron outside the dot. In this case no electron can tunnel onto the dot and no current flows. We can increase the energy of the electrons by imposing a voltage difference across the dot. Once the voltage reaches a level sufficient to accelerate electrons to an energy greater than the Coulomb blockade, the probability for tunnelling is dramatically enhanced and current can flow.

The essential step which turns this device into a transistor is adding a component that enables the charge on the dot to be controlled independently, thus giving a three-terminal device. The Coulomb blockade appears as a

Figure 4.5   Principles of the Coulomb blockade

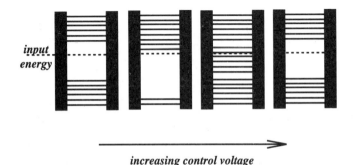

increasing control voltage

Note: An illustration of how the Coulomb blockade works. Inside a quantum dot there are a set of allowed energies, represented by solid black bars. The large gap in the ladder of allowed energies represents a Coulomb blockade. If the electrons were not charged particles, this gap would not exist. But because electrons are charged, some energies are excluded. If the input energy of an electron (dashed line) happens to fall in the gap, the electron cannot tunnel onto the quantum dot. By providing an external control of the voltage on the dot, we can remove the gap for certain voltages, thus enabling an electron to tunnel into and out of the dot. In this way a small bias can control quantum tunnelling through the dot and thus the current through the dot. This is just the kind of control needed for transistor action, but here it takes place at the level of a single electron.

gap in the allowed energy bands inside the dot (see Figure 4.5). However, by changing the charge on the dot it is possible to eliminate the Coulomb blockade. The reason for this is that now the energy of the electrons on the dot is determined not just by the capacitance of the isolated dot, but also the electrostatic energy of the charge in the dot and the third bias lead. When an electron is transferred to the dot the capacitative energy goes up, but if the initial charge on the dot is just right, the electrostatic energy goes down by exactly the same amount. With the Coulomb blockade gone, there are many allowed energies that the electron can tunnel to inside the well and thus the current increases. In effect, a small bias on the third lead can control a tunnelling current through the dot. This is just the kind of control required for transistor action, but in this device charge is controlled at the level of a single electron. By combining quantum tunnelling with charge discretisation we can build a single electron transistor.

The single electron transistor will probably become a usable device in the not too distant future. In typical devices the capacitance is not so small that the Coulomb blockade cannot be overcome by hot electrons. Thus it is necessary to cool the device to very low temperatures. To get an SET operating at room temperature will require even smaller quantum dots to be manufactured.

A slight modification of the basic principle of the SET yields a device that produces such a current so regular that it may be adopted as a quantum current standard. This device is called the electron turnstile. Next electron please! The first realisation of such a device in a semiconductor quantum dot was achieved by the Delft-Philips group in the Netherlands. In this extraordinary device split-gate techniques are used to periodically raise and lower the barriers separating the dot from the 2DEG (see Figure 4.6). The raising and lowering period is carefully controlled to a precise frequency. In the first step the left barrier is lowered just sufficiently to permit a a single electron to tunnel to the dot. The device is small enough that the

Figure 4.6   The electron turnstile

(a)   *(b)*

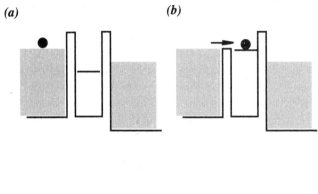

(c)   *(d)*

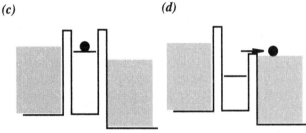

Note: A quantum dot, together with the Coulomb blockade effect, can be used to make an electron turnstile. In this device the barriers isolating the dot from the 2DEG are periodically raised and lowered to allow a single electron to be moved through the dot at regular intervals. No more than one electron can tunnel onto the dot at each step because of the Coulomb blockade. In (a) the barriers are too high to enable tunnelling. One barrier is then lowered (b) to enable an electron to tunnel onto the dot. Once an electron is on the dot electrostatic forces prevent further electrons from tunnelling through —the Coulomb blockade effect. The first barrier is then raised (c) and the second barrier lowered (d) to enable the electron to tunnel out of the dot. The result of this is a highly regular current in which single electrons are clocked through the dot at a regular rate.

Coulomb blockade now sets in to prevent another electron from tunnelling onto the dot. The first barrier is raised again and the second barrier is lowered to enable the electron to tunnel out of the dot. The process is then repeated. The device thus periodically clocks a single electron through the dot very much like a strictly controlled turnstile. The lowering and raising of the barriers is achieved by a frequency modulation of the gate potential which forms the dot and the entrance and exit barriers. The current flowing through the circuit is simply one electron per period of the applied gate frequency. The Coulomb blockade is responsible for the incredible stability of this current, which may ultimately be adopted as a new current standard.

In 1993 Hitachi Europe's Cambridge laboratory used the Coulomb blockade effect to construct a memory cell which in principle can operate at the level of a single electron. This device, if made to operate successfully at room temperature, would have an enormous commercial impact. Currently the best available commercial memories (DRAM chips) used the charge stored on a capacitor to represent a stored binary number. A number of clever techniques have been employed to keep the physical size of these capacitors as small as possible to enable more and more to be packed onto a single chip. Typically such a capacitor stores about 200 000 electrons. If it becomes necessary to rewrite the data this charge must be dissipated. The heat so produced is proportional to the number of stored electrons. A 16MBit chip dissipates about a tenth of a Watt. If the same chip was scaled up to store one terabit (TBit), representing one million times more data, the amount of heat dissipated might destroy the device. (Such issues are discussed in more detail in chapter six.) Clearly a memory element based on storing a single electron offers great potential as computer technology slowly moves to the terabit range.

The single electron transistor (SET) and a single electron memory represent the ultimate limit of downsizing in microelectronics. They are possible simply because the

Figure 4.7 Electron interferometers by coupled quantum dots

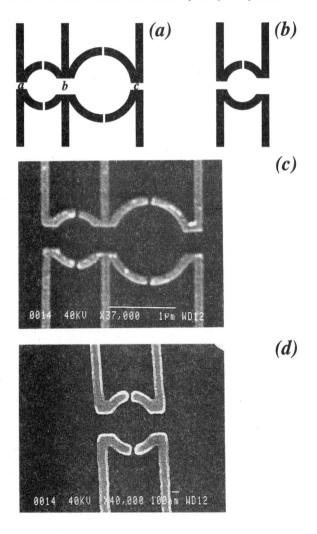

Note: A representation of the pattern of surface split-gates used by Taylor to create a quantum interference current device. In (a) there are six gates which can be individually controlled to form depletion regions in a 2DEG lying underneath. The small gaps in the top and bottom will pinch-off in the 2DEG depletion region. The gaps a, b, c can be left open to form a series of three QPCs. The result is two quantum dots joined by QPCs. In (b) the bias on the right most two gates is turned off, so that only one quantum dot forms in the 2DEG. Also shown in (c) and (d) are scanning electron microscope (SEM) photographs of the devices corresponding to these schematics. These devices were made by Richard Taylor and co-workers as part of the collaboration between the University of New South Wales in Australia and the National Research Council in Canada.

---

electronic charge comes in discrete packets. This is not strictly a quantum effect in the sense I discussed in chapter one. The basic idea of a Coulomb blockade which lies behind these devices can be explained entirely in terms of a small single charged particle every bit like a charged billiard ball. But such devices usually make use of quantum effects such as tunnelling. Recent experimental results suggest it may be possible to create an entirely new class of electronic device that directly exploits the strange rules of the quantum and interfering probability amplitudes.

QUIT

In the SET, quantum effects such as quantum tunnelling are combined with the natural discretisation of charge to enable interesting device possibilities. The crucial innovation which makes such devices possible is the ability to build circuit components with very small spatial extent, thus yielding very small capacitances. Single electron devices do not exploit one of the essential features of quantum mechanics: the ability of probability amplitudes to interfere with and cancel each other for different physical configurations of electron motion. If we could directly manipulate these probability amplitudes in a circuit we would be able to control currents by the direct cancellation of probability amplitudes. Such control would form the basis of a totally new, and far grander challenge, the quantum interference transistor or QUIT (the semiconductor community's predilection for acronyms is rather unfortunate in this case.)

Quantum interference devices hold promise for very fast switching controlled by tiny voltage swings. Both the Cavendish Laboratory group and the Delft-Philips group have demonstrated quantum interference control in a 2DEG with surface-gate techniques. I will describe instead some of the work undertaken by Richard Taylor at NRC in Canada and now with the University of New South Wales in Australia.

One of Taylor's quantum dot circuits is shown in Figure 4.7. In the figure there are six surface-gates. When a bias voltage is applied to the triplets, top and bottom, the effective barrier becomes continuous, and the result is two small pools for electrons connected by three, in-line, quantum point contacts (marked a, b, c in the figure). This device was fabricated in 1991 by Richard Taylor and co-workers at NRC in Ottawa, Canada. The innovative design, based on six gates rather than two, meant that various parts of the barrier could be turned off, allowing the study of transmission through a single dot, a double dot or various other combinations. The radii of the two quantum pools are 250 nanometres and 450 nanometres. The width of the three QPCs connecting the dots to each other and the 2DEG outside, is about 200 nanometres. A nanometre is about the same size as ten atoms placed end to end . . . these are very small constructions. Taylor also introduced a magnetic field perpendicular to the plane of the 2DEG. The magnetic field gives further control of the motion of the electrons. Magnetic fields give us a whole new ball game.

Electrons moving in a plane perpendicular to an external magnetic field will move in circles, unless they hit a wall. That is hardly likely in a bulk metallic conductor or

Figure 4.8   Edge states

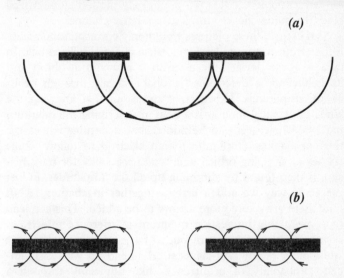

(a)

(b)

Note: If a magnetic field is directed perpendicular to the plane of the 2DEG, electrons moving in the plane are forced to move in circles, until they hit a boundary. The result is a kind of 'curvilinear' billiard game. If the magnetic strength is low, as in the top figure, the radii of the circular motion is large. This makes it difficult for electrons to get though a QPC. If the magnetic strength is high enough, electrons 'skip' along the walls, and can be guided through the channel.

semiconductor. But in the highly constrained motion of an electron in a 2DEG with surface-gate imposed boundaries, hitting the wall is par for the course, especially if the magnetic field is very strong. The kind of thing that can happen near a wall or QPC is depicted in Figure 4.8: electrons entering the channel skip along a wall. Figure 4.8a shows what happens to electrons trying to enter a point contact for a weak magnetic field. The radius of the magnetically imposed circular orbits is larger than the width of the constriction, and electrons are easily deflected. In

Figure 4.8b we see what happens near the point contact for large magnetic fields. Now skipping motion along a wall tends to guide the electrons around the channel.

Of course, these pictures would only correspond to reality if the electron were indeed a classical charged billiard ball. In fact, in these devices the electron is a quantum object and its behaviour is determined probabilistically through probability amplitudes. However, it turns out that knowing the kinds of classical motion is a help in describing the quantum motion. What needs to be added is a determination of the total probability amplitude for an electron to follow whole classes of skipping orbits. The total probability for transmission is then found by summing up all the amplitudes, in just the same way we added arrows together in chapter 1, but now there are many more arrows to be added. This can lead to some highly interesting quantum interference effects, as certain classical orbits turn out to have very low probability while other highly non-classical orbits end up with a significant probability of occurrence (like tunnelling through a classically forbidden region). Skipping motion in a classical picture then corresponds to so-called 'edge states' in a quantum picture in which transmission probability is largely determined by motion along a wall.

Apart from the imposed magnetic field, the Taylor experiment was similar to that used to demonstrate quantised conductance discussed earlier. A voltage was applied across the 2DEG outside the quantum dots and electrons injected into the first QPC. Electrons emerging from the final QPC at the second quantum dot then form a current, from which the resistance of the circuit, for the imposed voltage, can be determined. Consider first the case of a single quantum dot (that is a bias voltage is applied to only four surface-gates to define a quantum dot). At low magnetic fields, electrons entering the dot skip around the edges, but unless the magnetic field is just right, they will skip right over the output QPC and not leave the dot at all, resulting in a high resistance. As the magnetic field is changed, the geometry will be just right to ensure the skipping electrons entering the first QPC skip straight into the throat of the exit QPC. Electrons leave the dot, resulting in a higher current and thus a lower resistance. This is a kind of billiard game, but one in which the balls follow curved paths not straight lines. If we plot the resistance versus applied magnetic field we see peaks and troughs in the resistance. A peak corresponds to the call, 'electron to back pocket'. (What a pity someone can't make a billiard game in which the balls did follow curved lines, like the electron in a magnetic field. It might at least make televised pool games rather more interesting. Here is one way to do it: play billiards on a merry-go-round.)

These simple predictions are borne out by experiment, but there is nothing particularly quantum about them. They are easily explained without resorting to quantum probability amplitudes. However, at very high magnetic field and very low temperatures a true quantum effect takes the stage. If the magnetic force is great enough, the electrons form into wall-hugging loops around the interior edges of the dot. If the bias applied to the surface-gates is carefully tuned, it is possible to ensure that a barrier is raised just high enough to prevent such a looping electron from leaving the dot. The barrier is not so high as to prevent many other electrons, with different energies, from passing through the QPC. The trapped electrons will circle all the way around the dot, forming closed loops.

The presence of these loop states can change the path of an electron trying to pass through the dot. An electron can change from a through state into a loop state and then back into a though state, without any change in its overall energy. This means there are many possible paths that an electron can follow in passing through the QPC, including some in which the electron enters the dot from one side and is returned to the same side, i.e. it is reflected by the QPC. To find the total probability for an electron to pass through the QPC we need to determine the total probability amplitude for each of these paths and add them all up. In

117

some cases, as we shall see, cancellation can occur so that the probability of a loop state is zero.

Varying the magnetic field passing through the dot changes the probability of finding electrons in trapped orbits. What effect does all this have on the current? If we remove the possibility of a trapped orbit, electrons have no choice but to pass through the dot, giving a current larger than would be the case if some electrons became trapped. Varying the magnetic strength thus leads to a rise and fall in the resistance as trapped orbits are first more probable and then less probable. The reason that we can control the probability of trapped orbits this way is due to a new way of manipulating probability amplitudes that only arises for charged particles. It is called the Aharonov-Bohm effect after the two physicists, Yakir Aharonov and David Bohm, who first drew attention to this fact. To see what this new handle on probability amplitudes is we need to step back a bit in time.

The great synthesis of nineteenth-century research in electric and magnetic forces, provided by James Clerk Maxwell in the early 1860s, showed that the motion of charges is entirely determined by the strength of electric and magnetic fields. In providing this synthesis Maxwell showed how the fundamental forces could be rewritten in terms of potentials, an apparent mathematical artifice for succinctly representing the possible forces that may arise for a given arrangement of charges and electric currents. In many cases the potentials might be non-zero in a region where there were no electric and magnetic forces. However, as only the magnetic and electric forces can move a charged particle, the physical status of the ghost-like potentials was regarded with some suspicion.

With the discovery of quantum mechanics the mysterious potentials of Maxwell took on a new significance. It soon became apparent that the quantum probability amplitudes of a charged particle could be changed by the potentials alone, even in regions where there were negligible electric and magnetic forces. To observe these changes it

Figure 4.9   The Aharonov–Bohm effect

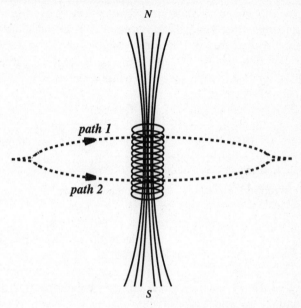

Note: In the Aharonov–Bohm effect electrons follow two different paths around a region containing a high magnetic field. The magnetic field in this case is formed by a current-carrying loop and is thus confined to the cylindrical region inside the loop. Thus no magnetic force is exerted on the electrons passing outside. Classical physics indicates that, because the electrons feel no magnetic force, they are not affected. Quantum physics, however, indicates that the probability amplitudes for electrons moving along the two paths are rotated in opposite directions. When the electrons are recombined it is possible that the probability amplitudes will cancel precisely. Even though the magnetic force is zero in the region through which the electrons move, there is a non-zero electromagnetic potential. It is this potential which acts directly on the probability amplitudes. The potential gives us another way to control probability amplitudes for charged particles, even if no force acts on those particles.

is necessary to look for situations where the outcome depends on combining a number of probability amplitudes. If the electromagnetic potentials change the amplitudes, they can change the conditions required for cancellation of probability amplitudes. Thus the electromagnetic potentials give us a new handle to control the probability amplitudes, at least for charged particles—a handle that does not exist for classical physics.

The full significance of this fact only became apparent after the pioneering work of Yakir Aharonov and David Bohm in 1959. They proposed an interesting thought experiment in which electrons could travel along two different paths around either side of a long current-carrying loop wrapped around a cylinder (see Figure 4.9). Inside the loop the current induces a strong magnetic field along the axis of the cylinder but outside there is almost no magnetic field at all, and the electrons should experience no force. Aharonov and Bohm noted, however, that the potentials outside the cylinder are not zero at all and can cause different rotations of the probability amplitudes depending on which side of the cylinder the electrons pass. Suppose we turn off the magnetic field. Now electrons travel along identical paths either side of the loop and thus the probability amplitudes will rotate by precisely the same amount from one side to the other. But when a current passes through the solenoid, the probability arrows for each electron get slightly out of phase and thus when recombined can give partial or even total cancellation. This would give a smaller probability to detect an electron at the point where the electron paths come together again. By changing the strength of the current through the wire we can vary the amount by which the probability amplitudes get out of step and thus look for oscillations in the detection rate for electrons where the beams meet. Of course it is not easy to get electrons to behave like this in free space, but it is precisely what electrons are forced to do in Taylor's device.

In Taylor's device, some electrons are forced by the magnetic field and the confining potentials to move in loops inside the dot. A loop can correspond to two kinds of motion, clockwise rotation or anticlockwise rotation. In a classical world these would have the same energy and be equally probable. But electrons in Taylor's device are quantum particles in a magnetic field and thus we need to determine the probability amplitudes for each kind of loop state. Because of the Aharonov–Bohm effect it is possible to arrange the magnetic strength so that the amplitudes for counter-rotating loops cancel exactly. In which case there is no chance at all of an electron entering into a loop. All it can do is pass straight through the dot. For other values of the magnetic strength the amplitudes add exactly and there is a high probability for a loop state to form at an energy available to the electron. In that case it can be tossed back from the QPC by passing through the loop in the right way. The result is an oscillation in the resistance of the dot as the magnetic field is varied.

Harnessing this effect to make a practical device took a big step forward in 1994 in an experiment undertaken in a collaboration between the University of North Carolina and IBM Research Laboratories in Yorktown Heights. In this case the confining structure was not based on GaAs but rather was fabricated in a silicon/silicon-germanium heterostructure. This potentially makes the device compatible with standard silicon-based semiconductor devices used in commercial microelectronics. Currently the temperature requirements are not as easily achieved as for GaAs.

It is now clear that quantum interference can be controlled to produce a new kind of microelectronic circuit. Current devices only operate in the laboratory at low temperature. It will be a big challenge to bring the QUIT to the marketplace. Advances in fabrication at nanoscopic scales should take us further and further in that direction. A huge investment is already being made in new nanofabrication systems. In Australia the Semiconductor

Nanofabrication Facility is currently being built in Sydney with a primary objective to build Si/Ge nanostructures. This facility, and others like it around the world, will take us towards a future based on quantum technology.

CHAPTER FIVE

QUANTUM CRYPTOGRAPHY

For the most part, the global communication system, constructed at enormous effort and cost, is used to send digital garbage, random strings of binary digits. This may sound like a serious underutilisation. On the contrary, it merely indicates that many messages are of a confidential nature. Currently the standard way to encode messages to prevent eavesdropping is to hide them in a torrent of random numbers. However all such schemes suffer from a fatal flaw; if an eavesdropper gets hold of the random numbers used, they can crack the code. Worse still, according to classical mechanics, they can do this in such a way that they will leave no footprint. In a world governed by quantum rules this need not be so. Quantum mechanics is an inexhaustible supply of randomness. But it is a randomness of a very special kind, so special that any eavesdropper, no matter how clever, must leave a footprint. In this chapter I will describe the new schemes for quantum cryptography, which can't be cracked and which can't be secretly eavesdropped (Brassard, Bennett and Ekert 1992).

Privacy and security in the digital age

History does not record the first use of a personal signature to authenticate a written document. The idea is no doubt as old as writing itself. A signature serves two purposes. It indicates to a reader the authenticity of the text and it prevents the author from repudiating the text at a later stage. The elaborate public ceremonies surrounding the signing of important international treaties attest to the almost sacred status of a signature. More prosaically, we all appreciate the importance of the signature on the back of our credit cards.

Yet as the world increasingly moves to electronic networks as the primary means for information exchange, the idea of a signed document will require some rethinking. Furthermore the openness of networks such as the Internet raises additional problems. Unless steps are taken to prevent it, an electronic document can easily be intercepted and altered for possibly criminal ends or perhaps as some kind of practical joke. This problem is particularly acute when the data available has monetary significance. Before I start shopping from this terminal, I want to know that my credit card number, once revealed to a vendor, is not going to be instantly available to every hacker on the net.

The need to ensure privacy in public communication channels is of course not new. Code making, and code breaking, is at least 2500 years old and has played a crucial part in the history of the world from the death of Mary Queen of Scots, to Enigma and the breaking of the German codes in the Second World War. The primary purpose of a code is to encrypt data in a form unreadable by all except the communicating parties. In addition to standard encryption, the exchange of electronic information has driven the need to develop new techniques for document authentication such as 'digital signatures'. The need to protect information on the electronic superhighway has turned cryptology into a billion-dollar technology, a technology which the US Government goes to a lot of trouble to keep from falling into the hands of so-called unfriendly states. Cryptographic devices, including software, are defined as munitions—the lessons of history have been heeded.

Traditionally, cryptography has been based on a secret key known only by the communicating parties. The key is used to both encrypt and decrypt the information. The first unbreakable code, using this method, was invented by Gilbert S. Vernam of American Telephone and Telegraph Company and Major Joseph O. Mauborgne of the US Army Signal Corps in 1917. Here is an example of the Vernam cipher. The letters of the alphabet are first coded as one- and two-digit decimal numbers based on a fixed rule known to the sender and receiver. If this were all there was to it the code could be easily cracked by searching for patterns. Now break the resulting string of numbers into five-digit blocks, and to each resulting five-digit number add, without carries, another five-figit number chosen at random, and transmit only the result of the addition. The set of random five-digit numbers is the key, and must be known by the sender and receiver. Anyone intercepting the message cannot decode the text unless they have the random key. This was the code used by Che Guevara, the Bolivian revolutionary, to communicate with Fidel Castro of Cuba.

At the time it was proposed, the Vernam cipher was not proved to be uncrackable. That proof would not come until the 1940s when Claude Shannon, the inventor of the mathematical theory of information (see chapter 6), proved that, so long as the random key was as long as the message being sent, the code could not be broken. This means we need to distribute very long random keys every time we need to send a reasonably sized message, a serious practical drawback. The key distribution problem represents the Achilles heel for this kind of code, for if any eavesdropper gets hold of the key, they can decode the message.

Nonetheless, a fixed-size, random key is satisfactory under some circumstances and forms the basis of the widely used data encryption standard (DES), originally developed by IBM and endorsed by the US Government as an official standard in 1977. The initial coding is not done with decimal digits but rather with the binary digits, 0 and 1, the universal alphabet used by computers. The principle remains the same with addition being done in base two. The DES uses a random key with 52 binary digits (that is a 52-bit string), which is used for many encryptions before being changed. DES is often implemented in hardware, making encryption very fast. Despite the fact that DES is widely available around the world, the US Government strictly regulates the export of DES either in hardware or software.

The DES code can be broken, precisely because the keys are of a fixed length, but it is far from easy. A head-on attack, searching all possible keys, would require 2^{55} steps on average. This would take many years on the very best general purpose computer (although a purpose-built machine might do better). Very recently, Eli Biham of Technion and Adi Shamir of the Weizmann Institute of Science, have devised a new technique that is much faster than exhaustive searching. While this is a theoretical milestone, it is believed to be impractical as it requires access to the DES device which did the encryption.

The late 1970s represented something of a watershed for cryptology, when Whitfield Diffie, Martin E. Hellman and Ralph C. Merkle, then at Stanford University, discovered the principle of public key cryptography (PKC) in 1976. Subsequently, in 1977, Ronald L. Rivest, Adi Shamir and Leonard M. Adleman, then at Massachusetts Institute of Technology, came up with a practical implementation of the idea known as RSA (Rivest, Shamir and Adleman). Public key cryptography overcomes the serious problem of secret key distribution which makes schemes like the Vernam cipher so vulnerable, as in PKC the communicating parties do not need to agree on a secret key beforehand. Even more useful in today's electronic information exchanges, PKC provides a route to digital signatures and document authentication. In public key cryptography documents are encrypted by a public key, known to the communicating parties and anyone else for that matter, as it is transmitted over public communication channels. The decryption, however, can only be done with a secret private key held by the receiver.

To explain PKC, let me introduce you to the central cast of modern cryptography, Bob and Alice and their nemesis, Eve. Bob and Alice are the names invariably given by crypto-theorists to the prototype communicating parties. Eve is, of course, the eavesdropper. Alice chooses randomly a pair of inverse mathematical operations. One operation can be used for encryption and one can be used for decryption. The particular mathematical operations involved (which for RSA require finding the prime factors of large numbers) have a trapdoor property—they are easy to do in one direction but not the other. Thus, knowing how to do the encryption is not good enough (or so we hope!) to be able to figure out the instructions for doing the decryption. Alice can broadcast her encryption instructions to everyone including Bob, who can now send a message that only Alice can decode. Likewise Bob can distribute his encryption algorithm so that anyone, including Alice, can encrypt a document that only he can read.

Alice can also transmit a document with her digital signature using PKC. To sign a message she does a computation involving the message and her private key to produce a file which she attaches to the message when it is sent. Steps must be taken to ensure the digital signature cannot be separated from the message and used nefariously. To verify the message, Bob does a calculation involving the message, Alice's digital signature and her public key. If the results of the calculation verify a simple mathematical relation, Bob can accept the document as genuinely from Alice. As Alice must closely guard her private key, she cannot later repudiate the document. Even better, digitally signed messages can be proved authentic to a third party, such as a judge, should Bob and Alice become adversaries in some legal wrangle. It would make a very interesting case, as the legality of a digital signature, and with it the future of the information superhighway, will ultimately be decided in court. Stay tuned, for you can be sure that this case is on its way. Pity the poor jury as they struggle with the mathematical subtleties of prime factoring and RSA!

Public key cryptography has not been proved to be secure. In fact in 1982 Shamir showed that an early PKC system, the knapsack cipher, could be cracked. PKC will only work if it is the case that no one can figure out the private key from the public key. If Eve does discover the private key she can not only intercept encrypted messages, she can forge signatures as well. The issue of the security of RSA reduces to the mathematical problem of finding factors. Providing Eve uses only classical computers, RSA is probably secure, as finding factors is thought (but not proved) to be a hard problem for a classical computer. In August 1977, when Martin Gardner published a description of RSA in *Scientific American* (Gardner 1977), Rivest, Shamir and Adleman issued a challenge to anyone who could factor a 129-digit number. At the time the possibility of anyone meeting the challenge seemed remote, but in 1994 Derek Atkins and Arjen Lenstra managed to find the factors. Certainly they had a lot more raw computing power available to them than was available in 1977 (they used the 'idle hours' on hundreds of workstations around the world over a period of almost a year; Brassard 1995). More significant was the availability of more sophisticated algorithms. Nonetheless, these advances in hardware and algorithms would still not suffice to factor a 500-digit number, so RSA can easily be made secure by using larger keys. However, as we shall see in the next chapter, if Eve has access to a quantum computer all bets are off! In that case Alice and Bob have a final resort—quantum cryptography.

Qubits and the Heisenberg conspiracy

All measurements are an attempt to crack the code in which is written the blueprint of nature. Quantum theory indicates that the universe is, at heart, an unbeatable random number generator, but rolled up in these numbers is the very order that gives us the world we see. The classical world view is based on the notion that the physical world is *out there*, that the plan is written for anyone to read. The quantum world view is quite different. The very measurements that we choose to read the plan predetermine which parts of the message we can read. Our instruments are as much a part of the quantum world as are the objects of the investigation. An instrument is like a private decoding key. It can only decode that part of the message corresponding to that particular key. The very act of measuring, of decoding if you like, recodes the message, thus changing future possible readings of the text. Rather than simply reading the blueprint, quantum theory tells us that our measuring instruments read and rewrite simultaneously. The physical principle which tells us how measuring instruments constrain the message they read is known as the Heisenberg uncertainty principle. It is this principle which makes it possible for Alice and Bob to communicate securely, immune from a surreptitious attack by Eve.

In chapter 1, I discussed a coin toss using light of very low intensity, so low that there were only one or two photons present. In the quantum version of a double coin toss we saw that the quantum probabilities give a result very different from the classical two-up game. The quantum odds for a simple experiment with only two outcomes can be quite different from the classical version.

In any experiment with two mutually exclusive outcomes, such as a coin toss, we can code the results in binary form. That is to say, we can represent the two distinct outcomes by the numbers 1 or 0. When the result of a fair coin toss is reported, information is gained. This elementary observation connects, somewhat paradoxically, random processes and the scientific measure of information formulated by Claude Shannon. The information acquired when one of two equally likely outcomes is realised is one bit of information. To store the result requires one binary digit of memory. Now we know that a quantum coin toss can be quite different from a classical coin toss. Does this mean that we need a new information theory to describe such experiments? How much information is acquired when the result of a quantum coin toss is reported?

Ben Schumaker, of Kenyon College in Gambier, Ohio,

has tried to emphasise the distinction between a classical coin toss and a quantum coin toss by introducing the notion of a qubit, the elementary amount of information acquired in a quantum coin toss. As we shall shortly see, there is a lot of physics hidden in the notion of a qubit, so much so in fact that it may be possible to use qubits to write unbreakable codes.

Qubits are what the world is made of, or so we think. Despite this they are not lying around in abundance; we need to go to a bit of trouble to manufacture them in a reliable way. A qubit, like a bit, describes an experiment with only two outcomes, which we will label 1 and 0. The essential characteristic of a qubit, however, is that the probability to get a 0 (or a 1) depends on *two* probability amplitudes. Thus we need to specify a total of *four* probability amplitudes to determine the odds of getting each result, 0 or 1. This may seem like an unnecessary complication. After all there are only two possible results to the experiment. Why not just toss a coin and be done with it? The reason qubits are more interesting than garden-variety bits is this: there are ways to manipulate each of the four amplitudes 'behind the scenes' and thus influence the odds of getting a 1 or a 0. The ability to manipulate nature at the level of probability amplitudes is the defining characteristic of quantum technology.

Here is one way to do it. A Mach-Zhender interferometer is a classical optical device. In its simplest form it consists of two perfect mirrors and two half-silvered mirrors (the kind I discussed in the quantum two-up experiment of Hong in chapter 1). A beam of light is split on the first half-silvered mirror. The two outgoing beams are then redirected by two perfectly reflecting mirrors towards another half-silvered mirror. The mirrors are carefully adjusted so that the two beams meet exactly on the second half-silvered mirror (see Figure 5.1).

Now inject a single photon into the first half-silvered mirror. This photon can either be reflected (R) or transmitted (T). Each of these occurs with a chance of 50 per

Figure 5.1   A Mach–Zhender interferometer

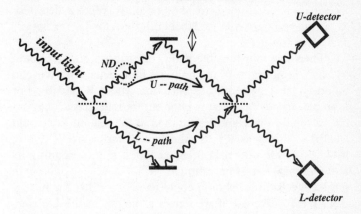

Note: A Mach-Zhender interferometer, when used with single photons, produces qubits. The input photon first encounters a half-silvered mirror where it is reflected or transmitted with equal probability. It can now follow two paths, the upper path (U) or the lower path (L) to the final beam splitter. Finally the single photon is detected at either the upper detector (U) or the lower detector (L). Two other variations are shown. Firstly, we can adjust the relative path length of the two branches by making small displacements to the upper mirror, as indicated by the double-headed arrow. Secondly, we could insert a non-demolition photon detector (ND) to see which particular path the photon took.

---

cent. In this experiment we have no way of knowing which possibility was realised, as we only detect photons after the second half-silvered mirror. Never mind, we will pretend that a photon was either reflected or transmitted at the first half-silvered mirror. Now the same choice has to be made at the second half-silvered mirror. If the photon was reflected at the first mirror, it can be either reflected or transmitted, with equal probability, at the second. Likewise, if it was transmitted at the first mirror, it still makes a

50/50 choice to be reflected (R) or transmitted (T) at the second mirror. Whatever happens, the only thing we know for sure is at which detector the photon was finally detected. Referring to figure 5.1, we label each detector U (for upper) or L (for lower).

There are four possible paths, or histories, that a photon takes in its journey from input to final detection. We can uniquely label each history with the letters R and T. For example, in the case that a photon was reflected at the first half-silvered mirror and transmitted at the second, we would label this history as RT. So the four histories are: RR, TT, RT, TR. But our measurement does not distinguish all these possibilities. For example, if we count a photon in the L-detector, that photon could have had either the history TR or RT. Likewise if we detect a photon in the U-detector, it could have got there by either of the paths RR or TT.

The classical probability description for this experiment is just the same as for a toss of two coins. Each path or history is equally probable and as there are four paths the probability of each is 1 in 4. But as we don't distinguish all four cases, but only pairs of cases, we add the odds for each case to get a 50/50 probability for detection in either output detector. This result depends only on the fact that a single photon has a 50 per cent chance of being reflected or transmitted at each of the half-silvered mirrors. It makes no reference to any other property of the device. For example, it should not matter if each of the two arms of the device were not the same length. Unfortunately this is not correct. The probability of a photon arriving at the U- or L-detector depends very strongly on the relative lengths of the two paths, even when there is no more than a single photon in the device at a time. If our classical particle picture is correct, the photon arrives at the last beam splitter entirely alone. Even if photons could interfere with each other in this particle picture, there is no photon for our lonely light quanta to interfere with. How can the path not taken have any influence on the result? Clearly our classical particle intuition has failed.

The correct description, the quantum description, is in terms of probability amplitudes (see Figure 5.2). Instead of assigning probabilities for each possible path or history, we assign probability amplitudes. If the experiment cannot distinguish two paths we must add the probability amplitudes for those paths, not the probabilities. Probability amplitudes, unlike probabilities, can be negative and thus can cancel when added together. So even though the individual probability amplitudes for the two ways a photon can get to a particular detector are not zero, the overall probability can be. In the example considered here we can set up the device so that the probability amplitudes for the histories RR and TT cancel exactly (Figure 5.2(b)), while the amplitudes for the histories RT and TR are identical and thus reinforce each other (Figure 5.2(a)). When this happens a photon is detected at the L-detector with certainty and never detected at the U-detector.

The results of this experiment are not random. As it is set up, a photon is detected at the L-detector every time. Despite this we cannot assign a single history to the result, but only the pair of histories RT and TR. In other words we can consistently assert that this photon suffered one reflection and one transmission.

The probability amplitudes for the photon histories in the device can be modified simply by changing the relative path length. This is easily done by moving one of the perfectly reflecting mirrors slightly. We can even arrange for none of the amplitudes to cancel so that the probability of detection is 50/50, just like the classical theory of probability would predict (see figure 5.2(c) and (d)). In fact by changing the probability amplitudes we can vary the probabilities continuously from 100/0 to 0/100 for detection in the U- or L-detectors respectively. We could set up the device so that the amplitudes for the histories RT and TR cancel exactly. Then we will be certain to detect a photon at the U-detector and can consistently assert

Figure 5.2   Quantum theory of a Mach–Zhender interferometer

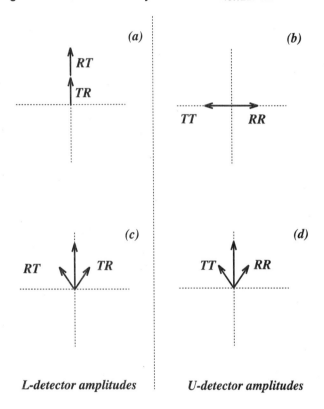

L-detector amplitudes　　　　*U-detector amplitudes*

Note: To determine the probability for detecting a photon at each of the two outputs (U and L) of the Mach-Zhender interferometer shown in Figure 5.1, we must add the probability amplitudes for the two histories leading to each detector. A history describes what happens at the two half-silvered mirrors. For example, in order for a photon to be detected at the upper detector (U) it must have either been reflected at both half-silvered mirrors (RR) or transmitted at both half-silvered mirrors (TT). In (a) we calculate the total amplitude to detect a photon in the L-detector. For this case the amplitudes point in the same direction, and the photon is certain to be counted at the

L-detector. Amplitudes for histories which determine a detection at the U-detector are rotated and cancel exactly (b), indicating we will never see a photon at this detector. By adjusting the relative path length, the amplitudes rotate. In each of (c) and (d) the relative path length is such that the amplitudes are rotated by 45 degrees. In that case photons are not transmitted or reflected with certainty.

---

that any photon we detect underwent either two reflections or two transmissions. More than this we cannot say.

The ability to control the probability amplitudes directly is the essential feature of quantum mechanics captured in the notion of a qubit. This control is the defining characteristic of a quantum technology.

The device I have described is a coin toss for photons which can now be biased by the turn of a knob. By controlling the path length we can bias the coin toss either to favour 'tails' (L-detector) or 'heads' (U-detector). We could have done this classically by using beam splitters that were not 50/50. However, quantum theory gives us a much more interesting way to bias the device by directly controlling the quantum probability amplitudes. Technically this kind of control is referred to as a *unitary transformation*, a concept at the heart of quantum computation discussed in chapter 6. If photons were not subject to quantum rules, but behaved like ordinary particles, the device with 50/50 beam splitters would always be a fair coin toss. But because of the way path length determines the probability amplitudes for detection at the L- or U-detectors directly, we have an extra handle on the outcome of this coin toss. The classical version, if it existed, would be a one-bit device. The actual quantum device produces a qubit.

The Mach-Zhender qubit machine is not a theoretical construct; it has actually been built. In 1988 Alain Aspect of Institut d'Optique, Orsay, built and operated the device at the one-photon level. The results were exactly as predicted by quantum mechanics.

135

An interesting variation of Alain Aspect's experiment illustrates the essential feature of quantum mechanics which enables quantum cryptography. Suppose we mount a device in each arm which registered a photon without destroying it (a so-called quantum non-demolition measurement). In each run of the experiment we would know which of the options, reflected (R) or transmitted (T), at the first half-silvered mirror was actually taken. In this case the possible histories a photon can have, prior to detection at the L- or U-detector, are reduced to just the two possibilities, RR and RT. The first of these (RR) results in detection at the U-detector, while the second (RT) results in detection at the L-detector. There is only one way, only one possible history, for detection at the L-detector and one history for detection at the U-detector, given a detection at the non-demolition detector. The probability to detect a photon at the L-detector is determined by the single probability amplitude for the history RT. There is no possibility of two amplitudes cancelling. In reference to Figure 5.2, we have removed the probability amplitude for the two histories TT and TR, and in so doing eliminated the possibility for amplitudes to cancel. As a result, the probability for detecting photons at each detector reverts to 50/50. By determining which particular history the photon started out on, we have altered the possibilities for future measurements!

In reality life is not quite so black and white. My theory group at the University of Queensland has been studying ways to do such non-demolition measurements for many years. Barry Sanders, now at Macquarie University in Sydney, and I considered the consequences of putting such a device in one arm of the Mach-Zhender interferometer. If the non-demolition device registers every photon that it can possibly register, the photons are detected at the two outputs with even odds. No variation of the path length makes the slightest difference. By choosing to determine which path the photon took we have lost all control over the quantum probability amplitudes. However, we might occasionally miss a photon in the quantum non-demolition detector. Then the truth is a strange mixture of interfering quantum histories and a classical coin toss. The more certain is the early history the photon took at the first choice, the more the device tends towards a classical one-bit, unbiased coin toss.

In the act of determining which path the photon took we are destroying the possibility of using a qubit. The act of observation has rewritten the future possibilities for observation. The future must be consistent with the past, regardless of what observations we care to make. In a quantum world, performing a particular measurement necessarily modifies the outcome of future measurements in an uncontrollable way. This is known as the uncertainty principle, but it is usually stated a little differently.

The uncertainty principle is central to quantum mechanics, so it is worth taking a bit of time to get this straight. Most presentations of the uncertainty principle give it in the form discovered by Werner Heisenberg in 1927. Heisenberg's version makes explicit reference to measurements of so-called complementary observables. If two physical quantities are complementary we cannot measure both with unlimited accuracy. We can measure one at a time as accurately as we like. However, an accurate measurement of one necessarily randomises the results we will get for the other. Measuring one of the two physical quantities has a drastic effect on the future observations of the other. The position and momentum of a particle are examples of such quantities.

Of course, measurements are complicated physical operations. Typically you would begin by looking though a catalogue of instruments and ordering the appropriate device, for example an accurate current meter. Then you have to connect the measurement apparatus to the system you are trying to measure. In doing this you have to be careful that you don't destroy the very effect you are seeking. For very accurate high-tech measurements the process can be quite complicated. For our purposes all measurements eventually reduce to asking simple yes–no questions, for example is the particle 'here' rather than 'there'.

To see the principle in action we can return to the single photon experiment with a Mach-Zhender interferometer. We would like to ask questions that have definite answers. If we set the path lengths just right we can ensure that the detection of photons will not be entirely random. For example, we can ensure that a photon will be detected, with certainty, at the L-detector. Under these circumstances we are certain that any detected photon had either of the two histories RT or TR. In effect we have asked the question, 'did a photon undergo one reflection and one transmission?' Every time we detect a photon the answer to this question is 'yes'. On the other hand we could have set up the device to give a definite answer to the question, 'did the photon undergo either two reflections or two transmissions?' (RR or TT). To make the measurement corresponding to this question we need to adjust the path lengths so that the amplitudes for the histories RT and TR cancel exactly, while those for RR and TT add. Then a photon will be detected with certainty at the U-detector.

It appears that simply by adjusting the paths we can get answers to questions involving pairs of histories. We can get a defininte answer to the question 'was the history either RT or TR?' By a simple adjustment of paths we can get a definite answer to the question 'was the history either RR or TT?' But suppose we want to get a definite answer to the question 'was the history either RT or RR?' This means we want to know if the photon was reflected at the first half-silvered mirror but don't care what happens at the second.

To keep track of the various pairs of histories and possible questions, a table is useful (see Figure 5.3). The table has two rows and two columns. The first column gives all histories in which the photon was reflected at the first half-silvered mirror, and thus follows the upper (U) path through the device. In the second column are the two

Figure 5.3   The principle of complementarity

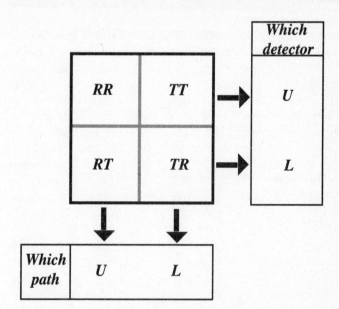

Note: The four possible histories we could assign to a detected photon are represented as a table. In the columns are the histories which describe a photon passing along different paths in the device. In the first column the photon history has a reflection (R) at the first half-silvered mirror and so must pass along the upper path (U). The rows contain the different histories leading to a count at the U-detector (first row) or the L-detector (second row). Depending on whether we read histories along rows or columns, we can ask 'which path?' or 'which detector?'

histories in which a photon was transmitted at the first half-silvered mirror, and thus follows the lower (L) path through the device. The column histories thus distinguish which path the photon follows through the device. The rows, on the other hand, correspond to histories leading to the different detectors at the output. The first row corresponds to all photons detected at the U-detector, while the second row correspond to the possible histories of a photon detected at the L-detector. The row histories thus do not distinguish what happened at the first mirror. As I now explain, questions (or equivalently, measurements) which distinguish 'which row' are complementry to questions which distinguish 'which column'. We have already seen how to measure row-type questions and get definite answers. By simply adjusting the path lengths we can ask, 'RT or TR?' or we can ask 'RR or TT?' The problem now is how to go about setting up the device to get definite answers to the column-type questions. Even better, can we ask a column-type question and a row-type question simultaneously and still get a definite answer to both? If this were possible we would know for sure which of the four possible histories the photon took. Unfortunately this is not possible precisely because these two types of questions are complementary.

How can we set up a measurement to answer column-type questions? This is easy. All we need do is put a non-demolition photon detector just after the first half-silvered mirror. Let us suppose we put the non-demolition detector to determine if the photon was reflected (see Figure 5.1). If we detect a photon we have the answer represented by the histories in the first column, that is RR or RT. How about setting up a measurement to simultaneously answer a row-type question and a column-type question? We can first carefully adjust the path lengths so that, without the non-demolition detector, we get a photon detected in the L-detector with certainty. This would indicate that all detected photons had either a RT history or a TR history. To determine which of these was actually the case we now insert the non-demolition detector into one arm of the device to see if a photon was reflected at the first half-silvered mirror. If this device works perfectly we get a definite answer if a photon was indeed reflected.

Now the problem is apparent. As previously discussed, the very act of making this observation renders the final detection completely random. No longer is it the case that a photon will be detected at the L-detector with certainty. No matter how we adjust the path lengths this situation will not change. A certain answer to a column-type measurement has rendered a row-type measurement completely uncertain. Row-type measurements and column-type measurements are complementary. In a classical world, we can expect a definite answer to the question, 'which row and which column?' In quantum mechanics this question does not have a definite answer because the physical properties joined by 'and' are complementary. The probabilities for different histories leading to a single observation are determined at a deeper level by probability amplitudes—that is the essential mystery of quantum mechanics.

The ability to manipulate the amplitudes directly opens up interesting possibilities for new technologies. One such possibility is the idea of using quantum randomness to encrypt data. This is quantum cryptography. Even more astonishing is the ability to use the quantum nature of the universe to develop a totally new kind of computational device, a quantum computer, which is the subject of the next chapter.

The quantum protocol

Some years before the invention of public key cryptography, quantum cryptography was born, but the world missed it. In 1970, Stephen J. Wiesner, then at Columbia University, discovered how to use quantum mechanics to combine two classical messages into one quantum message, from which the receiver could extract one message but not both. Wiesner also suggested a way to produce bank notes that would be physically impossible to forge. Wiesner's paper was rejected by the editors of the journal to which it was sent, and it remained undiscovered until 1983.

The second chance for quantum cryptography took place between 1982 and 1984 when the American physicist Charles H. Bennett of IBM Thomas J. Watson Research Center in New Jersey, working with the Canadian cryptographer Gilles Brassard from the University of Montreal, realised that qubits could be used to distribute a private key in such a way that no eavesdropper, no matter how careful, could monitor a signal undetected. Their method used a property of light know as polarisation, which is also subject to the Heisenberg uncertainty principle. In 1989 Artur Ekert and David Deutsch at Oxford University developed another protocol using correlated pairs of photons. The scheme I will describe here was invented by Charles Bennett in 1992, and uses the same Mach-Zhender interferometer discussed above.

The scheme is represented in figure 5.4. Alice uses a source which injects single photons into the right-hand port of the interferometer. Alice can change the path length by adjusting one of the perfectly reflecting mirrors on the left.

Figure 5.4   A qubit crytographic channel

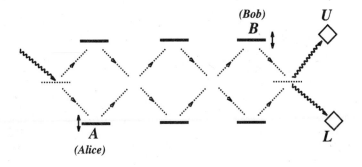

Note: A Mach-Zhender qubit machine configured as a quantum cryptographic channel. The communicating parties, Alice and Bob, can introduce path-length differences into each arm independently, as indicated by the double-headed arrows.

Figure 5.5   The quantum protocol

Alice's rotations	Bob's rotations	P(U)	P(L)
0 ←→ RT ⇑ TR RR TT	0 ←→ RT ⇑ TR RR TT	0	1
0 ←→ ⇑	45 ↘↗ ↘↗	1/2	1/2
45 ↘↗ ↘↗	0 ↘↗ ↘↗	1/2	1/2
45 ↘↗ ↘↗	45 ⇑ ←→	1	0
90 ⇑ ←→	0 ⇑ ←→	1	0
90 ⇑ ←→	45 ↘↗ ↗↘	1/2	1/2
135 ↘↗ ↗↘	0 ↘↗ ↗↘	1/2	1/2
135 ↘↗ ↗↘	45 ←→ ⇓	0	1

Note: A table showing the effect of the eight possible rotations of the probability amplitudes used by Alice and Bob on the final detection probabilities. In only four cases is the detection of a photon, at a single detector, certain. It is only these cases that Alice and Bob use in establishing a secure random key.

This will rotate the probablity amplitudes for the various histories leading to a detection at either output. Let us suppose that Alice randomly selects only four possible changes to the path-length leading to only four rotations: no rotation, a rotation by 45 degrees, a rotation by 90 degrees and finally a rotation by 135 degrees. These possible rotations are shown in Figure 5.5, for each of the four histories a photon encountering two half-silvered mirrors can follow. Alice chooses which length change to use in a completely random way, perhaps by tossing two coins. For example, if the coins both come up heads she may use a 45 degree rotation, or if both coins come up tails she may do nothing. What Alice is doing is randomly choosing four different probability amplitudes to describe the photon sent down the line to Bob.

At the other end of the device is the receiver, Bob. Bob can also change the path lengths by moving one of the perfectly reflecting mirrors on the right. Bob either does nothing, or changes the path length just enough to rotate the amplitudes by 45 degrees. We also assume that Bob decides what to do in a completely random way, perhaps by tossing a single coin. Now we need some way for Bob to assign a binary number to a sequence of observations. Let us assume that if the output photon is detected in the L-detector Bob assigns it the binary value 0. If it is detected in the U-detector, he assigns it the binary value 1 (see Figure 5.6).

The probability of getting a 1 or a 0 will depend on the actual rotations that Bob and Alice happen to use in a particular run. Suppose we set up the device so that if Alice and Bob happen to do nothing, a photon is detected in the L-detector with certainty. In this case the probability amplitudes for the two histories RR and TT cancel exactly. If Alice uses a 45-degree rotation while Bob does nothing, a photon is detected in either L- or the U-detector with equal probability. If Alice uses a 45-degree rotation and Bob also uses a 45-degree rotation, the net effect is to rotate the amplitudes by 90 degrees. Now the amplitudes for the histories, RR and TT, cancel exactly and a photon is detected in the U-detector with certainty. The various possibilities are summarised in Figure 5.6. There are eight possible combinations of rotations. In four cases the result

Figure 5.6   Bit assignment

A \ B	0	45
0	1	--
45	--	0
90	0	--
135	--	1

Note: A summary of how Alice and Bob assign binary digits once they have exchanged information about what kind of rotations Bob actually used. Only the cases in which detection at one detector or the other is certain are 'printed'.

---

is uncertain, that is a photon can be detected at L- or U-detector with equal probability. In the remaining four cases the result is certain, that is either detection at L- or U-detector with probability of one. In two of the certain cases the photon is detected at the L-detector so that Bob assigns the binary result 0. In the remaining two certain cases, the photon is detected at U-detector and Bob assigns the value 1.

Now we come to the protocol that Alice and Bob use to share a secret binary string for subsequent encoding and decoding. Firstly, they publicly agree to keep only those instances for which a photon is certain to be transmitted in either the U or L output. (Of course, at some point they will need to compare what rotations they used, in order to determine which instances to keep.) Alice and Bob agree to call this a *print* run, as only in this case will a 1 or 0 be recorded. Alice transmits a photon, randomly applying one of her four rotations. Bob then randomly chooses one of his two possible rotations, before finally detecting a photon and recording both the result, and which rotation he used (that is nothing or 45 degrees). The process is then repeated until a large number of photons have been sent and recorded.

Next Bob tells Alice, publicly, which rotation he used for each run, but does not tell in which output (U or L) the photon was detected. Alice then tells him which of his measurements corresponded to a *print* run. All other data is discarded, including those runs for which Bob failed to detect any photon at all (no photon detector is perfect). Alice and Bob now share a random binary string. The protocol is summarised in Figure 5.6.

The final step is for Alice and Bob to check for the presence or otherwise of Eve, the eavesdropper. Eve must make measurements on the photons inside the interferometer. In a quantum world this can make uncertain an otherwise certain outcome to a final detection at the L- and U-detectors. This would mean that one of the *print* runs was actually a random run, so that Alice and Bob might have recorded different results but don't know it. To detect this, Alice and Bob must sacrifice some of their laboriously obtained data. Alice and Bob decide to compare the *parity* of a publicly agreed-on random subset of their data. The parity of a set of ones or zeros is even if it contains an even number of ones, otherwise it is odd. For example, Alice might check her subset and tell Bob that it included an even number of ones. If Bob now checks his subset and finds this not to be the case, he immediately concludes his data set is different from Alice's. It turns out that if Alice's and Bob's data sets differ, the parity check will discover this with a probability of 50 per cent. Repeating the check 20 times with 20 different random subsets reduces the chance of an undetected error to less than one in a million. If an error is detected, Eve may have been at work, in which case the entire data set is discarded and Alice and Bob will have to try again later.

If an eavesdropper is present then errors will *inevitably* occur. To see this we must consider what the eavesdropper, Eve, would need to do. Suppose Eve discovers what kind of photons Alice sent, that is, she knows what rotation Alice used. Later, she can listen in as Alice and Bob exchange information to determine what path length changes Bob is to regard as correct. She then has as much information as Alice and thus knows exactly what binary digit Alice and Bob have arrived at. Eve then has the key. This sounds like it might work, and it would if the photons Alice sends had the rotations written on them for all the world to read, just like classical properties may be read without disturbing them in any way. Unfortunately this is not the case in the quantum world. A probability is not a property of anything, and neither is a probability amplitude. What you measure is not a probability but some other physical quantity, such as the presence, or otherwise, of a photon. Of course, the probability amplitude determines the relative frequencies for the various results in a long sequence of measurements, but each individual measurement is not a measurement of a probability.

In order for Eve to determine what kind of photons Alice sent, she will have to tap into both channels of the interferometer. I will not explain in detail the problem faced by Eve (see Figure 5.7 if you are really interested). The fundamental reason Eve cannot fail to be detected is this: in order to discover what rotation Alice used she would need to measure two complementary physical properties simultaneously. Quantum mechanics means that she cannot

147

Figure 5.7   Quantum spying

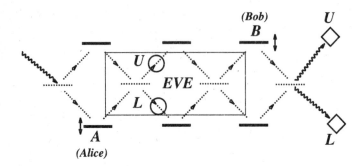

Note: An indication of how an eavesdropper, Eve, might intercept the cryptographic channel shown in Figure 5.4. Eve tries to determine the rotation used by Alice. To do this she must construct an identical interferometer to that used by Alice and Bob. The figure indicates how this may be done simply by inserting two additional half-silvered mirrors and two non-demolition photon detectors, U and L. To see the explanation more clearly I will only treat the cases which would lead to a *print* run, in the absence of Eve. The entire device is now simply two identical interferometers back-to-back.

Suppose neither Alice nor Bob uses a rotation. A photon will certainly pass Eve's L-detector. It then enters the next interferometer where it does exactly the same thing and emerges in Bob's L-detector. The first measurement, Eve's measurement, indicates that the photon had the history 'RT or TR', and the second measurement just confirms this. If Alice uses 90-degree rotation and Bob uses no rotation, a photon certainly passes Eve's U-detector and likewise passes Bob's U-detector. So far so good. But now suppose Alice uses a 45-degree rotation. To get a *print* run from this case Bob must use a 45-degree rotation. However now a photon can be detected at either of Eve's detectors with equal probability, but Eve can't know if each run was truly random. Suppose she counts the photon in the U-detector. This could equally well have occurred if Alice had used a 0-degree rotation. In any case what happens in Bob's detector must be consistent with Eve's observation. Bob's interferometer will behave just as described by the amplitudes in Figure 5.2(c) and (d), and thus a photon will emerge with even probability in each output detector. Instead of the outcome being certain as required for a print run, it is now random. This will result in Bob recording an error for about half of the print runs for this case. Eve can determine if Alice used a 0-degree and 90-degree rotations without introducing any error into Bob's results. However, if Alice uses a 45-degree or 135-degree rotation, sooner or later this will result in Bob recording a wrong result in a *print* run, and Eve can be detected. Of course, Eve can set her device up in such a way that she can determine 45- and 135-degree rotations and remain undetected, but now errors would be introduced whenever Alice used a 0-degree or 90-degree rotation, as you can confirm yourself.

---

do this without randomising one or the other of those quantities. The four rotations used by Alice can be partitioned into two sets of complementary histories for the photons that she sends to Bob. One set corresponds to rotations of 0 and 90 degrees. The other set corresponds to rotations of 45 and 135 degrees. Eve can set up a measurement to determine, with certainty, only two rotations at a time.

So here is Eve's problem. She can choose to monitor either rotations of 0 or 90 degrees and remain undetected or she can choose to monitor either 45-degree or 135-degree rotations and remain undetected, but she cannot simultaneously monitor both classes together and remain undetected. Alice has cleverly chosen to send qubits randomly chosen from two complementary classes subject to the Heisenberg uncertainty principle. No matter what Eve does she cannot measure both kinds of qubits with the one apparatus and remain undetected. She must leave her footprint somewhere.

In classical physics, we can code one bit of information in any physical property with two values, such as the different patterns on the faces of a coin. That information can simply be read off, without any changing of the physical system. In the quantum world a bit of information is encoded in the result of a measurement. Probability amplitudes determine the frequencies of the results. There is no further explanation in terms of classical properties. Once the measurement has been made, the probability amplitudes for future measurements will, in general, change. This is certainly true if we make measurements of two things restricted by an uncertainty principle. Quantum cryptography works because an eavesdropper is forced to make measurements subject to an uncertainty principle, which necessarily change the transmitted qubits uncontrollably.

Key distribution is secure in a quantum world provided your key is a qubit. However, sending qubits turns out to be a rather difficult matter. The problem is that qubits can be disturbed en route to detection by things other than an eavesdropper. It would be very difficult to distinguish such disturbances from a real eavesdropper. It might even be possible for an eavesdropper to hide in these other disturbances. Despite this problem, quantum key distribution is sufficiently robust for it to have already been done in the laboratory.

The first quantum cryptographic channel was built and operated by Charles Bennett and Gilles Brassard and their colleagues at IBM in 1989. The device used polarised light to encode the qubit and was capable of transmitting a secret key over a distance of about 30 centimetres. This scheme was subsequently refined using new devices and materials and can now operate over distances of a kilometre or more. In 1991 John Rarity and Paul Tapster at the British Defence Research Agency in Malvern, England, succeeded in sending a key over a much longer distance. Their scheme was based on a correlated photon protocol inspired by the correlated pair protocol of Artur Ekert. It has now been refined to the point where it operates over a distance of many kilometres.

Quantum key distribution is a physical reality, but it requires an entire laboratory of state-of-the-art laser equipment to make it work. It is doubtful that it will become a commercial possibility for some time, especially as standard public key cryptography is used so widely with such confidence. However, as you are about to learn, public key cryptography may be open to a new attack which can in principle be mounted by a quantum computer. If this becomes a reality, quantum mechanics will be both the executioner and saviour of public key cryptography.

CHAPTER 6

THE QUANTUM COMPUTER

The twentieth century is the century of information, and the computer. It is also the century of the quantum and in 1985 David Deutsch, a physicist at the University of Oxford, published a paper in which these two intellectual currents merged: what kind of computation can be done in a quantum universe? The answer to this question is only beginning to emerge and it is overturning our long-accepted paradigm of computation, a paradigm based squarely on classical physics. The result is likely to be computational power which makes today's most advanced supercomputer look like an abacus.

Information costs

A computer is no less a physical device than a car engine or electrical motor, and as such is subject to physical limitations. However, there is a deep and fundamental difference between an engine and a computer. Where an engine processes energy, a computer processes information. The physics of the nineteenth century was largely driven by the need to understand what restrictions physics placed on the ability of an engine to process energy. A century of thinking about this question produced thermodynamics, one of the cornerstones of physics.

At the close of the nineteenth century the entire universe was viewed as a gigantic machine, something like a steam engine, chugging along in accordance with the laws of thermodynamics. In the first decades of this century this view of the universe took a severe beating as the quantum revolution got under way. Quantum theory tells us that the universe is fundamentally and irreducibly random. A highly constrained and puzzling form of randomness to be sure, but random nonetheless.

It has taken us a long time to appreciate what quantum randomness really means. The quantum theory, our best theory of physical reality, is actually a theory, not of physical *things*, but of physical *information* (even today not every physicist would accept this point of view). Increasingly physicists are following the path suggested by John Wheeler (1989): 'Tomorrow we will have learned to understand and express *all* of physics in the language of information'. It is perhaps only in the closing stages of the information century that information could be conceived of as having a reality, just as in the nineteenth century energy came to have a kind of reality.

At first sight a theory about probability and randomness would appear to be the complete antithesis of a theory of information. That randomness and information are intimately connected was elucidated in the work of Claude Shannon of Bell Laboratories in 1948. This is not an easy idea to get across. We are so used to the everyday connotation of the word 'information' as something that has meaning. Shannon's great insight was to disconnect the concepts of meaning and information. Consider the single toss of an unbiased coin. We expect, based on our initial ignorance, that heads and tails are equally likely outcomes. If the coin is tossed and the result is hidden from us, our ignorance is just as great as it was before the toss. When we learn the result, our ignorance is reduced as we acquire *information*. How can we quantify the amount of information so acquired? What would be the desired properties of a measure of information? These were precisely the questions answered by Shannon.

Let us now try to guess a measure of information. The

information gained in the result of a single yes–no experiment, like a coin toss, should be the smallest quantity of information. If we toss two coins, and learn the result, one expects intuitively that we have gained twice as much information as for a single coin toss. So information should be additive. Now suppose we knew the coin had two heads. When we learn the result of this coin toss, we have gained no information at all, as we already knew what the outcome would be. So information must always be a relative quantity, relative to whatever background information we already have.

Now suppose the coin did indeed have two sides but we knew it to be biased to come up heads more often than tails. This situation is intermediate between the fair coin toss and the toss of a double-headed coin, and we expect that the information acquired should be less than that for a fair coin toss. In this way we see that information is determined in some way by the probability of a given outcome. If the probabilities are equal, the information is a maximum. If one probability is one (and the other zero) as in a double-sided coin, the information is zero. All other cases lie in between.

We need not worry about the precise definition of information arrived at by Shannon. The basic ideas discussed in the last paragraph will be sufficient. Let me summarise. The maximum average information occurs when one of a set of equally likely outcomes is realised. Average information is additive and it depends on the probability we assign to a set of outcomes. We will take as our fundamental measure of information the average information obtained in the toss of an unbiased coin. In a language which should be rather appealing to modern computer-literate sensitivities, we will say that the average information acquired in the toss of a single unbiased coin is *one bit* of information. Then the average information in the toss of two unbiased coins is two bits of information, and so on. If we toss 8 x 1024 coins, the average information is 1024 bytes or, in modern computer jargon, 1Kb.

My use of computer jargon in the previous paragraph was not gratuitous. There is an intimate connection between probability, information and computer storage of numbers. Computers are essentially a large interconnected array of switches. Each switch can be on or off, two mutually exclusive possibilities, like the outcome of a coin toss. We do not represent the state of a computer switch as a head or tail but rather use the on/off state to code for the numbers 1 and 0. This basic alphabet is sufficient to represent numbers, letters, symbols, in fact anything at all. Earlier this century telegraph messages could be coded in a similar two-symbol alphabet called Morse code. The 1/0 code used in computers is called the binary number system. All the data and all the programs in your computer are coded as enormous strings of binary symbols. Binary strings are the lingua franca of computers and information.

Suppose we toss an unbiased coin a very large number of times. We will record the result of a head as 1 and a tail as 0. The final result is a very long string of ones and zeros, a long binary string. How much memory do we need to store a typical result? Computers store numbers coded in the form of strings of ones and zeros. In this case, then, the result of all our coin tosses is already in a form the computer can handle. It turns out that for a very large number of coin tosses, we need to set aside about as many places to store ones and zeros as the average information, as measured by Shannon's formula, for the large number of coin tosses. For example, if we toss 8192 coins we will need to set aside roughly 1Kb of memory. Some special results can be stored in a much smaller amount of memory. For example, it is possible that we will get all heads. Then instead of storing a very large string of ones we just store a single 1 and the binary code for the number of coin tosses. For a large number of tosses, this will require much less than the average information in the sequence of tosses. However, this result is very unlikely. The lesson here is that we must not focus on particular results, but phrase our description in terms of typical sequences and thus the average information is the appropriate measure of how much memory we need to set aside. It must also be held in mind that I am only referring to very large sequences.

In the example of the last paragraph, a sequence of all heads was seen to be particularly easy to store in a computer. A sequence of all heads looks distinctly suspicious as a typical result from a large number of coin tosses. We instinctively suspect that it is not obtained by the random process of coin tossing. The result seems just too regular and non-random. This intuitive notion has been refined to give a definition of randomness. Gregory Chaitin of IBM Laboratories and others have sought to define randomness in terms of the minimum amount of memory we need to set aside to store a number. This is the algorithmic definition of randomness. We saw above one example of this, in which we only had to store a one and the number of times it was to appear. But other examples are possible. Suppose we start writing down all the numbers appearing in the decimal expansion for π. Here is the expansion to 100 decimal places:

3.1415926535897932384626433832795028841971693993751058209749445923078164062862089986280348253421170 68

In many ways the resulting sequence is a random sequence of decimal digits. However, according to the algorithmic definition the number is not random. All we have to do is write a simple program to start computing the digits of π (simple algorithms exist for this) and the number of digits we want. Our supposed random number is now stored as a program code, occupying only a few bytes of memory, and a single number representing the number of digits we want. If it is your responsibility to buy new hard discs for your computer this is a very important saving.

At first sight this seems to make the idea of randomness highly dependent on the kind of programming language we use and indeed on the kind of computer we use. We can overcome the second objection by referring the definition to a Universal computer. This is not a real device (although it could be) but a set of rules for manipulating strings of ones and zeros. It comes as something of a surprise to learn that only a small set of rules is needed to be able to carry out all possible computations. This result was discovered by the British mathematician Allen Turing in 1936. In a real computer many rules, built out of a set of simple rules, are used to increase the speed with which the device can operate. The reason a Turing machine is called universal is that it can be used to simulate any other computer, albeit rather slowly (Hopcroft 1984).

To overcome the second objection we must take very seriously the idea that a computer program represents a kind of code. Using the theorems of coding theory Rudiger Schack, now at Royal Holloway College, London, has shown that the definition of algorithmic information is indeed quite precise, and numerically determined by Shannon's average information.

We see that information can be measured and has a real cost in terms of the computer memory needed to store a typical result of a trial. The final step in recognising information as a true physical concept came when Rolf Landauer of IBM, while thinking about the ultimate physical limits to computation, realised that to erase information costs energy. The result, now known as Landauer's principle, provides the link between the nineteenth-century concepts of energy and thermodynamics and the twentieth-century concepts of information. It is the crucial step which leads us to a physics of information. Landauer's principle states that to erase one bit of information requires that we give up a small amount of energy as heat, dissipated into an environment (Bennett and Landauer 1985). The exact amount of energy lost is proportional to the temperature of the environment. At room temperature, the actual amount is very small, roughly the kinetic energy of a single air molecule. Through Landauer's principle, information acquires a physical character and a real physical and economic cost. I will return to a more detailed discussion of Landauer's principle later in this chapter. For now let us reflect a while on the astonishing progress of computers and ask, can it continue?

The 3G-pc

The progress of computer technology in the last thirty years is unprecedented. No technology has gone further in terms of speed, efficiency and affordability than the computer. In this time the number of transistors on a chip has increased by a factor of one million. At the same time the price per transistor has decreased by a factor of 100000 and the power consumption has been reduced by a factor of 100000 as well. An interesting point of reference for these kinds of numbers has been given by Colin Warwick and co-workers at AT&T in New Jersey. They equate the price of a car with a computer and liken the capacity of a car to the memory of a computer, and the speed of a car to the number of instructions a computer can carry out per second. If improvements in cars had matched the computer industry since 1946, a modern car would weigh 60 grams, cost $40, have 1.5 million litres of luggage space, use only one litre per 600000 kilometres to travel at 2160000 kilometres per hour.

At a conference in early 1995 engineers described a memory chip that can hold a billion bits of information, and a microprocessor that performs more than one billion instructions per second. In technical language one billion is referred to by the preface giga. Thus a 3G-pc would have one gigabyte of memory, have a clock speed of one gigahertz and process one billion operations per second. Don't expect to be able to buy one for your desktop soon, wait a few more years. Engineers do not expect progress to stop there, but somewhere down the track, probably by the end of the first decade of the next century, a true limit will be reached, a limit imposed by the laws of physics.

Two very important problems need to be overcome as more and more transistors are packed onto a silicon chip.

Firstly, the size of each individual device must get smaller, as must the wires connecting them. Secondly, the heat generated by billions of little switches needs to be removed from the chip before it melts.

To understand the first of these problems we must consider how microprocessor circuits are made. The circuits on the chips are laid down by a photographic process and subsequent etching. The smallest dimension currently achieved in commercial devices is about 0.35 micron in certain memory devices. A micron is about the width of a coil of DNA or one-thousandth the width of a human hair. However, to reach gigabyte chips the dimension must shrink to about 0.1 micron, the so-called 'point one' goal (Stix 1995). At this point we reach the limits of the photographic etching process. Circuit patterns begin to blur and the light gets absorbed before it reaches the surface. It may be possible to overcome this problem, perhaps by using X-rays instead of ultraviolet light as currently used, but it is far from clear if such processes are economically viable in a factory setting. Not only must each transistor itself get smaller as the device density goes up, the small metal wires that connect devices on the chip must also shrink. A typical wire must withstand a current density roughly 200 times greater than the maximum current density in household wiring. These large currents damage or destroy very thin wires. In addition to this problem, the wires will be getting close together and 'cross-talk' becomes a problem.

The second difficulty is power dissipation, a real killer. The transistors that switch on and off in computer circuits are very inefficient and a great deal of power is lost as heat. Modern processors give off about as much heat as a cooking surface of comparable size. Heat is also generated even if the device is not actively switching. This leads to two problems. Firstly, if a great deal of electrical power is lost doing nothing very useful, the battery of a portable pc will run flat sooner. The rapid growth in portable computing provides a great incentive to find ways to minimise heat loss. But the real reason why engineers try to minimise heat generation is that, for very small circuits, it is very difficult to remove heat from the chip before it destroys the device. Literally, the device stews in its own juice. These problems indicate that the current technological and economic limits will be approached simultaneously for chips of the order of 3 square centimetres containing about one billion elements, with minimum features about 0.1 micron in size. These limits will be reached sometime in the next decade. But are these real limits? Are they required by the laws of physics, or merely the limits of a particular technology? Perhaps there is another way.

Reversible computation

In the 1960s Rolf Landauer of IBM Thomas J. Watson Research Center, New York, began to wonder if the heat associated with logical processing was required by the laws of physics or simply an historical accident—we just happened to have built computers out of inefficient transistors. The result of his thinking led to Landauer's principle, the fundamental result which connects abstract information theory with physics. But perhaps more importantly it stimulated a line of research which has led directly to the idea of a quantum computer, a machine that exploits the quantum rules, opening up a vista of computing power way beyond the limits of current technology.

Landauer's principle (Bennett and Landauer 1985) states that whenever a bit of information is erased, a small amount of energy must be given up as heat (roughly equivalent to the kinetic energy of an air molecule at room temperature). How do we erase information? Return to the example of a coin toss. Recall if the two sides of the coin are distinguishable we gain one bit of information when we learn the result of a single toss. If both sides of the coin were the same we would gain no information upon tossing the coin. Thus to erase information we must arrange for two previously distinguishable alternatives to become indistinguishable.

Here is another example. Suppose we have a small box containing a single air molecule. We can turn this device into a casino game by asking on what side of the box is this molecule located—is it on the left or the right? The molecule is zipping around inside the box so fast that the result of successive observations should be a random sequence of results, left or right. Clearly this is a molecular coin toss. To destroy information we have to arrange for the molecule to be found only on one side of the box. We can do this by pushing a piston in from one side until it has halved the volume of the box. If we push the piston in from the left the air molecule is now necessarily located on the right. We gain no information when we observe which side the molecule is on as we already know it must be on the right.

Now the model of a gas in a piston is exactly the kind of thing nineteenth-century physicists considered in their formulation of the laws of thermodynamics. The essential result we need is this: when a gas is compressed it must give off heat in order to remain at the same temperature. We can now consult a text on thermodynamics to see what is the minimum amount of heat we need to dissipate to halve the volume of a box containing one air molecule. The result is exactly the amount required by Landauer's principle for the erasure cost of one bit of information.

Every existing computer necessarily wastes energy. This is not just because engineers are careless, but is determined by the fundamental logical processing elements in the machine. Take for instance the AND gate. This is a device with two inputs and one output. If both input voltages correspond to the binary symbol 1, the voltage at the output corresponds to the symbol 1 also. For every other input, that is for all the other pairs of ones and zeros present at the input, the output voltage codes for a 0. Such a device is destroying information. For example, there are three different inputs, (1,0), (0,1) and (0,0) which all get mapped to the same output. As we saw earlier, this compression of distinguishable possibilities must destroy information and, by Landauer's principle, must be accompanied by a small amount of energy as heat. The kind of logic embodied in the AND gate is called irreversible logic, as we cannot reverse the output and obtain the input precisely because different inputs are mapped onto the same outputs. Landauer's principle shows that the device is irreversible in the thermodynamics sense as well, as potentially usable energy is lost as heat.

But must we use irreversible logic gates to build computers? The answer given by a number of physicists in the late seventies and early eighties is no. It is easy to see how to build a simple kind of reversible gate. Whenever information is processed by a gate we can store the input together with the output. Obviously it is always possible to logically reverse the operation as we already have the inputs. Needless to say the memory cost of a reversible computer built this way is going to be considerable.

A more efficient way to proceed is to search for new logical gates, alternatives to the AND gate, which are always reversible. It should be easy to see that such a gate must have the same number of inputs as outputs, so the question becomes, what is the minimum number of inputs a gate must possess in order to be sufficiently powerful to build all possible logical operations from it? The answer is, three. A number of such gates have been proposed. Here I will discuss the Fredkin gate invented by Ed Fredkin, now at Boston University.

A Fredkin gate has three inputs and thus, to be reversible, must have three outputs. One input is called the control and is left unchanged by the gate. That is, if the bit on the control input is 1 the bit at the output of the control is also 1. However, the logical status of the control line can change what happens for the other two inputs. If the bit on the control line is 0, the other two inputs pass through the gate unchanged. If the bit on the control is 1, the bits on the other two lines are interchanged. Such a gate is universal for computation in as much as every logical operation can be performed by a suitable network of such gates.

A computer built from reversible gates is itself reversible and so in principle need dissipate no energy whatsoever. However, as the computation proceeds the device becomes cluttered with bits left over from intermediate steps, bits that must be present if each operation is to be reversible. Somehow we must get rid of these bits if we are to reuse the computer. We cannot simply erase the trash bits, as that will immediately invoke an energy retribution via Landauer's principle. The answer was provided by Charles Bennett of IBM. We first make sure we copy the answer we want when the computer produces it (which can be done reversibly), then we just run the entire computation backwards, thus resetting all the internal logic states to their initial configuration. As each step in the machine is reversible there is no problem in reversing the computer. At the end of this we have the answer we want and the computer is always returned to the same state ready for the next calculation. The entire computer is a gigantic piece of plumbing which pipes a pattern of ones and zeros from the input to a different pattern of ones and zeros at the output. Of course, the pattern at the output is of interest to us as the result of the computation.

Currently computers consume vastly more (one million times more, to be precise) than the Landauer minimum per logical operation. A great deal can be done with conventional methods to lower the energy cost per operation without resorting to reversible logic. However, technologies are steadily approaching the Landauer limit. If present trends continue it is estimated that reversible technologies could become attractive by around 2020.

Several groups have realised that reversible logic can be implemented by conventional CMOS (complementary metal oxide semiconductor) technology. One group at University of Southern California has been developing prototype VLSI (very large scale integration) chips that use adiabatic switching, a kind of reversible logic, to reduce energy dissipation. The basic idea is to recycle circuit energies rather than lose them as heat. In such circuits the operating power is derived from the clock signals which serve both to power and synchronise the circuits. A similar approach is being used by John Denker at AT&T Bell laboratories in New Jersey.

Many of the seminal ideas in the field of reversible computation were presented at the First Conference on the Physics of Computation held at the Massachusetts Institute of Technology in 1981. A browse through the published proceedings indicates the themes that would dominate the discussion of the physical limits of computation over the next decade. Included in the proceedings are a number of papers that attempt to discover what limits quantum mechanics might place on a computer. Thus began the next phase of the discussion which paved the way for Deutsch's crucial contribution in 1985.

Quantum computers

Once it was realised that computers can avoid the Landauer cost by utilising reversible logic, it was time for quantum considerations to enter the picture. Ultimately any reversible physical system must be subject to the laws of quantum mechanics. Irreversible systems are not so interesting from a quantum point of view as, by definition, such systems are open to the environment, into which they can dump heat energy. Such interactions with the external world change the rules of quantum mechanics and mask such essential features as Feynman's rule (see chapter 1), the principle that tells us to add probability amplitudes rather than probabilities themselves.

I think the first person to suggest that quantum computers might be vastly more powerful than classical computers was the famous CALTECH physicist Richard Feynman. In the First Conference on the Physics of Computation held at MIT in 1981, Feynman asked whether or not the behaviour of every physical system can be simulated by a computer, taking no more time than the physical system itself takes to produce the observed behaviour. Is this not simply a question of making a sufficiently powerful and fast computer? Apparently not. The question really gets interesting when we imagine the physical system in question to be a quantum system. All computers available in 1981 (and now for that matter) were certainly not quantum systems. Feynman suggested that it may not be possible to simulate a quantum system in real time by a classical computer whereas it may be possible with a quantum computer. Predicting the behaviour of a physical system is just a particular kind of calculation. So if Feynman's suggestion is correct it implies there are tasks that a quantum computer can perform far more efficiently than a classical computer.

The first question to ask is, does quantum mechanics permit reversible computation? Might not the uncertainty principle somehow restrict our ability to do strictly reversible computation? Early investigations concentrated on showing how Turing machines and reversible gates, such as the Fredkin gate, could indeed be implemented in quantum mechanics. In 1988 I proposed a simple optical device, based on the Mach-Zhender interferometer discussed in the last chapter, that would implement a Fredkin gate at the level of single photons. (A similar idea was proposed independently and at about the same time by Yoshi Yamamoto, then at NTT in Japan.) The lesson was clear: as reversible logic was necessarily carried out by physically reversible devices, they would work just as well quantum mechanically as classically.

These early ideas sought merely to replicate a classical reversible gate using quantum objects. The whole discussion entered an entirely new domain when David Deutsch showed in 1985 that by directly manipulating probability amplitudes quantum reversible logic gates can do much more than their classical counterparts, thus fully justifying Feynman's extraordinary insight. It is now clear that a universal quantum computer can be built out of a gate with only two quantum inputs.

The idea of directly manipulating probability amplitudes to do computation has since been elaborated by a number of people, culminating in an astonishing result obtained in 1994 by Peter Shor of AT&T, New Jersey: quantum computers can factor large numbers very much faster than any classical computer. This result is of considerable commercial and military interest as one of the most common data encryption systems is based on the supposed computational difficulty of factoring large numbers.

Why should quantum computers be so much more powerful than classical computers? The answer lies in the notion of a qubit introduced in the previous chapter. A qubit is an elementary quantum process with only two, mutually exclusive, outcomes. The outcomes can be coded by the binary digits 1 or 0, or simply as heads or tails. However, unlike a coin toss, the probability of a particular outcome for a qubit is determined at a deeper level by a probability amplitude. Quantum theory shows us how to manipulate these amplitudes directly (a unitary transformation), and it is this fact that enables quantum computers to do much more than a classical computer.

To see how a quantum computer can do better than a classical computer we need to build one—at least on paper. Like any computer we start with elementary building blocks, called 'gates'. These building blocks are assembled into larger functional units, which then form the components for even larger units, and so on up to the final computer. I cannot take you all the way—no one really knows how to do that yet—but I can show you the elementary building blocks and a couple of simple constructions which will give you the general idea. I will begin with a simple example, first devised by David Deutsch and Richard Jozsa, called 'the square root of NOT'.

The square root of NOT

A NOT gate is simply a device for turning a head into a tail, or a 1 into a 0. That is, a NOT gate is a single input device which reverses the logical status of a single bit. The square-root-of-NOT gate is itself a single input gate. Two such gates run into each other give a NOT gate overall. So, in a manner of speaking, the square of a square-root-of-NOT gate is a NOT gate. Seems obvious enough, but how to do it?

In fact we have already done it. In the previous chapter I discussed a qubit-producing device based on a Mach-Zhender interferometer. This elementary optical device is illustrated schematically in Figure 5.1. The device has two inputs and two outputs, which I distinguish by the labels U for upper and L for lower. Light enters either the upper input (U) or the lower input (L) on the left. I will assume throughout that we are using light of such a low intensity that, in any run of the experiment, no more than one photon is injected into the device. Of course, that photon can go into either the U- or the L-input. Likewise we can only detect a single photon at either the U-output or the L-output. Whether we detect a photon at the U- or the L-output depends on the relative lengths between the two paths through the interferometer. I will label these paths also by the symbols U (for the upper path) and L (for the lower path).

There are two histories leading to a photon being detected in the U-output—the photon can either be reflected (RR) at both half-silvered mirrors or transmitted (TT) at both half-silvered mirrors. As the detection does not distinguish these two histories, the probability amplitude to detect a photon at the U-output is the sum of the probability for each history separately. Likewise for the L-output, with the two histories symbolised by RT or TR (see Figures 5.2 and 5.3). It is possible to vary the angle between the probability amplitudes for various histories simply by changing the relative path length. For example, if the path lengths are the same, the amplitudes for RR and TT may be identical, while those for RT and TR point in opposite directions and thus cancel exactly. In this case the photon is detected with certainty in the U-output.

In Figure 6.1, however, I illustrate a case where the amplitudes neither exactly cancel nor exactly reinforce. In this case the photon is detected with equal probability at either the U- or the L-output. This is precisely the case we need to construct a square-root-of-NOT gate.

It might look as though the device in Figure 6.1 is simply an elaborate coin toss, a simple one-bit game. In fact the device produces a qubit which behaves very differently from a coin toss. To see this, suppose we take the output of this device and run it straight into another identical Mach-Zhender interferometer, without looking to see which output the photon emerged in (see Figure 6.2). This causes the probability amplitudes to continue to rotate. The result is that at the final output, the amplitudes for the RR and TT histories cancel exactly while those of the RT and TR reinforce each other. The result is that the photon emerges with certainty in the L-output. Thus a photon at the U-input

Figure 6.1   A square-root-of-NOT gate

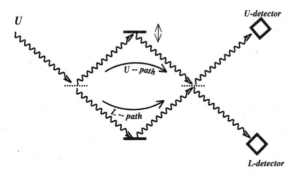

Note: A square-root-of-NOT gate (SRN) is a Mach-Zhender interferometer operating on a single photon at a time. The path lengths are adjusted so that the probability amplitudes to emerge in the U or the L channel have the same size. Then there is an equal chance to detect the photon at either channel. It might help to refer to Figure 5.2.

Figure 6.2   A quantum NOT gate

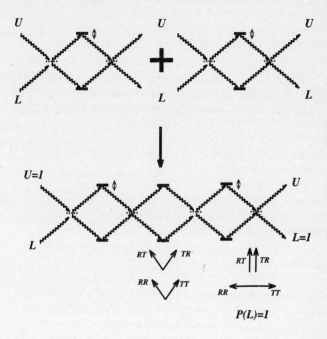

Note: Two Mach-Zhender interferometers, with unequal arm lengths, run back to back form a NOT gate. Each Mach-Zhender interferometer is then a square-root-of-NOT gate. This device only works because the random switching at each beam splitter takes place at the level of probability amplitudes rather than actual probabilities. The first interferometer thus produces a qubit rather than a coin toss. The second interferometer then manipulates the probability amplitudes from the first interferometer in such a way that the photon finally exits in the L-output channel. If each interferometer functioned according to the rules of classical probability, the photon would exit in both outputs with equal probability.

has been sent to the L-output. This is very different from what we would get if we simply injected photons at random into the two inputs of a single interferometer. In that case they would simply emerge at random from the final outputs. Because we have not determined which histories were actually realised at the first device, we must keep working with probability amplitudes which simply continue to rotate in the second interferometer.

Overall this device inverts the input. A U-input goes to an L-output and an L-input goes to a U-output. To make this a NOT gate we change our symbols to the conventional binary symbols 1 and 0. If we map U to 1 and L to 0, we see that the total device takes a 1 to a 0 and a 0 to a 1 (see Figure 6.2). This is a NOT gate. To summarise, two Mach-Zhender interferometers run end to end realise a NOT gate. With an obvious, if rather cumbersome, notation we can then refer to a single Mach-Zhender interferometer as a Square-Root-of-Not gate, which I shall simply refer to as an SRN gate.

In contrast, if the first Mach-Zhender device were simply using a coin toss to assign photons to the outputs, the same would hold for the second device and the photon could emerge in either output direction with equal probability. The whole device would not be a NOT gate but rather an elaborate coin toss game.

The interferometer realisation of an SRN gate is the elementary building unit from which we can construct a quantum computer. It has one input and one output so it is obviously reversible. However, it is a little too elementary to do much more than manipulate a single qubit. To make a quantum computer we need to manipulate many qubits. Fortunately a discovery made by David di Vicenzo at IBM, Yorktown Heights, and Adriano Barenco and co-workers in the Oxford quantum computing group in collaboration with Richard Jozsa at the University of Plymouth, shows that any computation on a register of qubits can be broken up into a series of operations on pairs of qubits. One way to do this is with a *controlled NOT* (CN) gate. I will show you how to construct a CN gate from a collection of SRN gates.

To do this we will need four SRN gates. We run two SRNs into each other to make two single bit gates. This would produce two independent NOT gates. We could go on with this, but it is simpler to readjust the path lengths in each SRN so that instead of two NOT gates we get two IDENTITY gates. (A NOT-NOT gate in fact!) The trick to making a CN gate is, not surprisingly, to get one gate to control the other.

To proceed we need to be clear about what kind of control we want. Two independent IDENTITY gates have two inputs and two outputs. Let us call the input to one of the gates a *control* qubit, and the other qubit the *target* qubit. Now we need to connect the two gates to realise the following rule: if the bit on the control qubit is a '1', invert the output of the target leaving the control unchanged, otherwise do nothing (see Figure 6.3).

The key to realising how to do this is to see that a controlled NOT gate is nothing more than a quantum non-demolition (QND) measurement of the state of the control qubit. It is a measurement in the following sense. If we see the target qubit become inverted we are sure that the bit on the control qubit was a '1'. It is a quantum non-demolition measurement because, whatever happens, the control qubit does not change. This is a very important point which does not seem to be widely recognised. Indeed every reversible quantum computation can be regarded as a QND measurement of the inputs. It is QND, because in order to be reversible we must ensure that the input is left unchanged. It is a measurement because the output is a function of the input. This immediately suggests that there is a close connection between the theory of quantum computability and quantum measurement theory. For example, is the class of measurable quantities larger than the class of computable functions?

A CN gate constructed from four Mach-Zhender interferometers is shown in Figure 6.4. Two pairs of

171

Mach-Zhender interferometers are coupled in series to produce two parallel IDENTITY gates. We will now couple these two gates together in such a way that the target gate performs a QND measurement of which path, upper (U) or lower (L), that the photon took in passing between the two component interferometers of the control gate. We will choose the coupling so that if we send a photon into the U-input of the target gate and it comes out in the L-input,

Figure 6.3   Truth table for a controlled-NOT gate

Input		output	
Target	*Control*	*Target*	*Control*
1	*0*	*1*	*0*
0	*0*	*0*	*0*
1	*1*	*0*	*1*
0	*1*	*1*	*1*

Note: A table summarising the principle of a controlled NOT (CN) gate. The device has two inputs and two outputs. One input is the *control* and the other is the *target*. If the control is in a '1' state, the bit on the target is reversed, otherwise nothing happens. Note that the bit on the control is always unchanged.

Figure 6.4 Principles of a quantum controlled-NOT gate

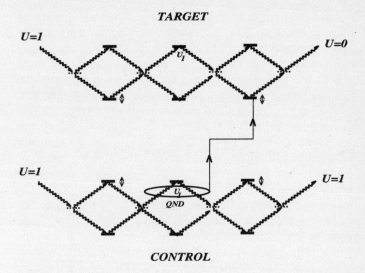

Note: An illustration of how to build a CN gate from four Mach-Zhender interferometers. The interferometers are connected to form two parallel gates each composed of two interferometers connected in series. The top series combination is the 'target' gate while the bottom is the 'control' gate. The path lengths in the interferometers are adjusted so that each gate acts independently as an IDENTITY gate. That is, we ensure that the first interferometer of each pair is adjusted so that a photon incident in $U=1$ emerges with certainty in the upper channel U_1 in the middle of the device, and the second interferometer in each pair is adjusted to ensure that this photon will emerge with certainty in the upper output channel. Now a QND apparatus is inserted into the upper channel U_2 in the control gate. If the QND device determines that a photon has indeed passed this way, it reversibly changes the path length in the target gate so that the photon in this gate is switched to the lower target output.

we are sure that the photon in the control gate passed along the upper path in the middle.

How do we couple the two gates together? It must be done in such a way that a photon passing through the upper part of the control gate can change the probability amplitudes in the target gate. In other words, we must ensure that if a photon passes along the upper path in the control gate it changes the path length in the target gate to such an extent that the amplitudes for the target photon to emerge in the U-output cancel exactly. There are suggested schemes to do this, none of which are very practical at this stage, but in principle it can be arranged. The first problem is that single photons don't make much of an impact on the world. You may think that this could be overcome with a suitable arrangement of amplifiers but there is a problem. The connection between the target gate and the control gate must be entirely reversible. If not, the whole device is no longer reversible and it cannot operate at the level of probability amplitudes. Classical devices such as linear amplifiers are necessarily irreversible. The connection between the control and the target gate must remain at the quantum level and the result of the QND measurement must remain suspended between potentiality and actuality. Fortunately there are easier ways to make a CN gate than by using photons.

The world's first quantum CN gate was demonstrated at NIST in Boulder, Colorado, in mid-1995 using a laser-cooled ion in a magnetic trap. In this experiment we see one aspect of a developing quantum technology, ion and atom trapping (see chapter 3), enabling another. (This kind of synergy is to be expected and suggests that any nation that wants to get ahead in the quantum technology race would be well advised to put together a fairly interdisciplinary quantum technology laboratory!)

The key to making a quantum computer is to be able to adjust all the probability amplitudes in just the right way, so that the probability amplitudes for a right answer all add up to give a large overall probability for being right. In addition to being able to exploit the unusual way probability amplitudes add, a quantum computer has another interesting feature—it can simultaneously do many calculations at once, simply by preparing the input as a superposition of all possible input states. Now that is interesting.

Beat the market with a quantum computer

The square-root-of-NOT gate takes a single bit, say 1, and produces an output which is a qubit or, in more technical terms, a quantum superposition of 1 and 0. If we make a measurement to determine which output is realised (if the bits are represented by photons we just count photons), we get 1 or 0 randomly with equal probability. This looks like a straight coin toss, but we know it isn't for the reasons discussed in the last section. That is, if this superposition is sent through another square-root-of-NOT gate, the final output is the inverse of the original input, whereas if the gates were just randomly switching bits the final output would still be a coin toss. We distinguish the kind of output produced by a single square-root-of-NOT from a coin toss by saying that the output is a coherent superposition of 0 and 1 rather than a mixture of 0 and 1. A coherent superposition of 0 and 1 is neither one state nor the other, it is only potentially a 0 or 1, with potentiality becoming an actuality upon measurement. Before measurement both possibilities are present simultaneously. Suppose we have a difficult decision to make that depends on one or the other of two mutually exclusive conditions having occurred. We can represent the decision process as a function of the binary inputs 0 or 1, which code for the two mutually exclusive input conditions to the decision. We could try and get a rational basis for our decision by first assuming the input is 1 and working the calculation through to a result. We then repeat the calculation again for an input of 0.

But suppose this is a very difficult decision with weighty consequences for error. For example, the decision may require making a crucial investment exactly 24 hours after the input conditions become available. If the fastest computer we have takes more than 24 hours just to run the calculation through for a 1 input alone, then we are sunk, and can do nothing. If we have a quantum computer, we can prepare the input in a coherent superposition of the two inputs and do both calculations simultaneously in 24 hours. Sure we may only get the right answer 50 per cent of the time, but we clean up every second day. Now you see the trick: a quantum computer can check out the values of a function on all possible inputs *simultaneously*!

The above example was given by David Deutsch in his 1985 paper. The details of how it is done are a little complicated. The essential idea is to allow the quantum computer to follow simultaneously all computational paths, transforming the probability amplitudes for each path in just the right way so that, when they are finally combined just before a readout, the required result is obtained with reasonable probability (and low or even zero probability of error). Since Deutsch's paper a number of people have used the idea to devise many interesting problems that can be solved more efficiently on a quantum computer than on a classical computer, culminating in 1994 with Peter Shor's factoring result. This has resulted in the new field of quantum computational complexity. It is now clear that a quantum computer would not just be a bizarre but useless device for astonishing classically minded physicists. It may be very useful indeed, so much so that perhaps we had better see if we can build one.

A number of groups have suggested physical systems that could be used to build a quantum computer. The discovery, by David Di Vicenzo, that we only need to manipulate two qubits at a time, rather than the three required for classical reversible computation, has made the search a lot easier. The optical gate proposed by my group at the University of Queensland and Yoshi Yamamoto is certainly a candidate as it can easily be run as a two-bit quantum gate. The crucial problem will be to get it to work with single photons. This requires large optical nonlinearities. Unfortunately optical nonlinearities tend not to be unitary state transformers due to their tendency to absorb light. Recent work by Jeff Kimble and his group at the California Institute of Technology, using very small cavities and single atoms, may overcome this problem.

Another very promising suggestion was made in 1994 by J. I. Cirac of Universidad de Castilla–La Mancha in Spain and Peter Zoller at the University of Innsbruck in Austria. The idea is based on the astonishing technology of atom optics described in chapter 3. Zoller and Cirac suggested using the quantised collective motion of a large number of ions in a trap to enable qubits to be exchanged, by a sequence of unitary transformations, between ions. The device would only be possible due to the extraordinary advances in laser cooling that were described in chapter 3. In this suggestion of Cirac and Zoller we see how new ideas in one area of quantum technology may yet lead to the ultimate realisation of a quantum machine, a quantum computer.

In mid-1995 Dave Wineland's group at NIST (National Institute of Standards Technology) in Boulder, Colorado, used the idea of Cirac and Zoller to build the world's first quantum CN gate. Of course this is not a quantum computer but it is the essential building block from which a quantum computer can be constructed. Wineland and co-workers are confident that their scheme can be extended to work on many qubits, not just two as in the CN gate.

Despite the huge surge of interest in the possibility of quantum computation, it is unlikely that any serious device will be available for many years. The basic problem is the need to ensure that all the quantum probability amplitudes are transformed in a completely reversible way (that is, unitarily). This requires that the computer be completely isolated from the outside world while the calculation takes place. A quantum superposition of only two states is notoriously fragile in the presence of any outside disturbance, and in a quantum computer we would need a

177

coherent superposition of thousands, even millions of qubits. Recent results in my group, and in other groups working on this problem, suggest that it may not be quite as bad as it looks. After all, a quantum computer usually only returns the right answer with some probability. It may be that if a few bits get kicked around by an unknown disturbance during the computation, it will not too seriously affect the final probability of getting a result. We are decades (at least) from realizing a quantum computer. If you are one of those people who simply must have the latest in computer technology on your desk (or in your lap), I am quite sure your decision will not be made any more difficult by the need to consider a quantum computer. Your grandchildren may have to face the choice. I expect, however, that a laboratory demonstration of the power of quantum computation is not far away. Perhaps in less than a decade the first purpose-built quantum factoring engine will be demonstrated. When that happens the ongoing arms race between code makers and code breakers will really heat up.

EPILOGUE

The exciting horizon

Technologies discussed in this book are but a sample of the emerging field of quantum technology.

These technologies reflect our increased understanding of the quantum world resulting from decades of fundamental research in physics. Some of the technologies, for example scanning tunnelling microscopy discussed in chapter two, are already in the marketplace. Others, such as quantum circuits described in chapter four, and atom optics in chapter three, will within a decade produce technologies occupying important niche markets. Quantum cryptography and quantum computing I suspect will become important on a rather longer time scale. These latter two innovations are still so new that it is very difficult to say just what, if any, will be the practical outcomes of this research. However one thing is certain, investigations in these areas will change forever the way we think of information and computation.

I am in no doubt that the future of high technology is quantum technology. The idea that the physical world can be manipulated at the level of the quantum is a powerful stimulus to the imagination of physicists and engineers the world over. It is hard to say just where these ideas will lead us. Thus far the possibilities are only being realised in distinct research communities. What we need is a greater dialogue between these communities and it might not be going too far to suggest that there is a place for a quantum technology laboratory in which various research directions discussed in this book could be brought together. The challenge of such a laboratory would be to bring quantum technology to the market. The potential for truly revolutionary innovation is enormous and so is the likely return on investment. Perhaps one day my grandchildren will be graduating from a Department of Quantum Engineering.

The world of the quantum may be bizarre, but it is our world and our future.

BIBLIOGRAPHY

Bate, Robert T., 1986, The quantum effect device: Tomorrow's transistor, *Scientific American*, March, p. 78

Bennett, Charles H. and Landauer, Rolf, 1985, The fundamental physical limits to computation. *Scientific American*, January, p. 38

Binnig, Gerd and Rohrer, Heinrich., 1985. The scanning tunneling microscope, *Scientific American*, August, p. 40

Binnig, Gerd and Rohrer, Heinrich, 1987, Scanning tunneling microscopy—from birth to adolescence, *Reviews of Modern Physics*, July, p. 615

Brassard, Gilles, 1995, The impending demise of rsa? *electronic source* http://www.rsa.com/rsalabs/faq/faq-rsa.html

Brassard, Gilles, Bennett, Charles H. and Ekert, Artur K., 1992, Quantum cryptography, *Scientific American*, October, p. 26

Braun, Ernest and Macdonald, Stuart, 1982, *Revolution in Miniature*, Cambridge University Press, Cambridge

Business Week, 1993, Windows on the world of atoms, 30 August, p. 48

Chu, Steven, 1992, Laser trapping of neutral particles, *Scientific American*, February, pp. 40–54

Cohen-Tannoudji, C. and Phillips, W., 1990, Vol *33* No. 10 New Mechanisms for Laser cooling, *Physics Today*, p. 33

Feynman, R. P., Leighton, R. B. and Sands, M., 1965, *Feynman Lectures in Physics, vol. 3*, Addison-Wesley, Menlo Park

Gardner, Martin, 1977, Mathematical games, *Scientific American*, August, p. 120

Hong, C. K., Ou, Z. Y. and Mandel, L., 1987, Measurement of sub-piosecond time intervals between two photons by interference, *Physical Review Letters*, 59:2044

Hopcraft, John E., 1984, Turing machines, *Scientific American*, May, p. 70

Kastner, Marc A., 1993, Artificial atoms, *Physics Today*, January, pp. 24–31

Mehra, Jagdish and Rechenberg, Helmut, 1982, *The historical development of quantum theory*, Springer-Verlag, New York

Sheehan, Danny and Lamotte, Wayne, 1985, *Heads and Tails; the story of the Kalgoorlie two-up school*, Uniquely Australian

Simpson, J. A. and Weiner, E. S. C., 1989, *The Oxford English dictionary*, Clarendon Press, New York

Stewart, I., 1989, *Does God Place Dice?* Basil Blackwell, Oxford

Stix, Gary, 1995, Toward point one, *Scientific American*, February, p. 90

Taylor, R. P., 1994, The role of surface-gate technology for algaas/gaas nanostructures, *Nanotechnology*, 5:183–98

Wheeler, J. A., 1989, Information, physics, quantum: the search for links. *Procd. 3rd International Symposium on the Foundations of Quantum Mechanics*, pp. 354–68

INDEX

183